50·00

C

Offshore Electrical Engineering

Offshore Electrical Engineering

G. T. Gerrard

Butterworth-Heinemann Ltd
Linacre House, Jordan Hill, Oxford OX2 8DP

 PART OF REED INTERNATIONAL BOOKS

OXFORD   LONDON   BOSTON
MUNICH   NEW DELHI   SINGAPORE   SYDNEY
TOKYO   TORONTO   WELLINGTON

First published 1992

British Library Cataloguing in Publication Data
Gerrard, G. T.
   Offshore electrical engineering.
   I. Title
   621.38209163

ISBN 0 7506 1140 5

Library of Congress Cataloging in Publication Data
Gerrard, G. T. (Geoff T.)
   Offshore electrical engineering/G. T. Gerrard.
   p.   cm.
   Includes bibliographical references and index.
   ISBN 0 7506 1140 5
   1. Electrical engineering.   2. Offshore structures—Electrical
   equipment.   I. Title.
   TK4015.G37   1992
   627′.98—dc20   91–31333
                                                            CIP

Typeset by TecSet Ltd, Wallington, Surrey
Printed and bound in Great Britain

Contents

Preface

I hope that many of the electrical engineering lessons which have been learned by experience during the last 20 years or so of North Sea oil endeavour are covered in this book in a way that people in all walks of life will find interesting.

Nevertheless, may I apologize in advance to those who may find the coverage lacking in some way. Please write to me if you have constructive comments to make and I promise I will bear them in mind when considering any future revision of the book. I cannot promise, however, to reply to all correspondence.

During the several years that this book has taken shape, many significant events have taken place within the oil industry; some have been good and others disastrous. If there is one precept worth adopting, it is the need to consider everything and every situation as new and unique. Be warned that blindly or rigidly applying regulations may cost lives.

Geoff Gerrard

Acknowledgements

The author wishes to thank and gratefully acknowledge all those who provided material and advice for the production of this book, particularly the following:

Stephen Rodgers, John Brown Engineering Ltd., Clydebank

Ian Stewart and Arlene Sutherland, BP Exploration Ltd., Aberdeen

Andrew White, Andrew Chalmers and Mitchell Ltd., Glasgow

David Bolt, Ewbank Preece Ltd., Aberdeen

Lynn Hutchinson, Ferranti Subsea Systems Ltd., Victoria Road, London W6

Hamish Ritchie, Geoff Stephens and John McLean, Foster Wheeler Wood Group Engineering Ltd., Aberdeen

Gordon Jones, G. E. C. Alsthom Large Machines Ltd., Rugby, Warwickshire

Mr P. G. Brade, G. E. C. Alsthom Measurements Ltd., Stafford

Mrs M. Hicks, Publicity Department, G. E. C. Alsthom Installation Equipment Ltd., Liverpool

Pat Dawson, Hawke Cable Glands Ltd., Ashton-under-Lyne, Lancs.

Gordon Shear, Hill Graham Controls Ltd., High Wycombe, Buckinghamshire

Richard Crawcour and Mr K. M. Hamilton, P & B Engineering Ltd., Crawley, Sussex

Sue Elfring, Crest Communications Ltd., for: Rolls-Royce Industrial and Marine Ltd., Ansty, Coventry

John Day, formerly with Shell UK Exploration and Production, Aberdeen

Jim Bridge and Keith Stiles, SPP Offshore Ltd., Reading

Ian Craig and Graham Sim, Sun Oil Britain Ltd., Aberdeen

Prof. John R. Smith, University of Strathclyde, Glasgow

Stephen Rodgers, John Brown Engineering Ltd, Clydebank

John Baker, GEC Alsthom Vacuum Equipment Ltd.

Mr John Hugill, Thorn Lighting Ltd., Borehamwood, Herts

Chapter 1

Introduction: offshore power requirements

Designing for provision of electrical power offshore involves practices similar to those likely to be adopted in onshore chemical plants and oil refineries. However, other aspects peculiar to offshore oil production platforms need to be recognized. It is suggested that those unfamiliar with offshore installations read the brief guide in Appendix A before continuing further.

The aspects which affect electrical design include the following:

1. The space limitations imposed by the structure, which add a three-dimensional quality to design problems, especially with such concerns as:
   (a) hazardous areas;
   (b) air intakes and exhausts of prime movers;
   (c) segregation of areas for fire and explosion protection;
   (d) avoidance of damage to equipment due to crane operations.
2. Weight limitations imposed by the structure, which require:
   (a) the careful choice of equipment and materials in order to save weight;
   (b) the avoidance of structurally damaging torques and vibrations from rotating equipment.
3. The inherent safety hazards presented by a high steel structure surrounded by sea. Such hazards often require:
   (a) particular attention to electrical shock protection in watery environments;
   (b) good lighting of open decks, stairways and the sea surrounding platform legs.
4. The corrosive marine environment.

1.1 Hazards offshore

1.1.1 Marine environment

Wave heights in the North Sea can exceed 20 metres, and wind speeds can exceed 100 knots.

1.1.2 Gas

Accumulations of combustible gas can occur on an offshore installation from various sources, including the following:

(a) equipment and operational failures such as rupture of a line, flameout of an installation flare, a gland leak etc.;
(b) gas compressor vibration causing failure of pipe flanges, loss of compressor seal oil etc.;
(c) drilling and workover activities;
(d) in concrete substructures, the buildup of toxic or flammable gases due to oil stored in caisson cells.

1.1.3 Crude oil and condensates

Equipment and operational failures, such as the rupture of a line or a gland leak, can release oil and condensates. The high pressures involved in some cases cause spontaneous ignition due to electrostatic effects.

1.1.4 Operational hazards

Apart from the fire and explosion hazard of process leaks, there is a hazard to personnel purely from the mechanical effects of the leak jet and the sudden pressure changes caused by serious leaks in enclosed compartments.

Care must be taken in the siting of switchrooms, generator sets and motor drives to minimize the risk of damage due to crane operations, especially if cranes are sited near drilling equipment areas where heavy pipes and casings are being frequently moved.

1.2   Electrical system design criteria

The purpose of any offshore electrical supply system is to generate and distribute electricity to the user such that:

1. Power is available continuously at all times that the user's equipment is required to operate.
2. The supply parameters are always within the range that the user's equipment can tolerate without damage, increased maintenance or loss of performance.
3. The cost per kilowatt hour (kWh) is not excessive, taking into consideration the logistical and environmental conditions in which generation and distribution are being effected.
4. Impracticable demands are not made on the particular offshore infrastructure, i.e. such as those for fuel or cooling medium.
5. The safety requirements pertinent to an offshore oil installation are complied with, in particular those associated with fire and explosion hazards.

6. The weight of the system is not excessive for the structure on which it is installed. In the case of rotating machinery, the effects of vibration and shock loads must be taken into account.

A single-line diagram of a typical offshore electrical system is shown in Figure 1.1.

1.3   Main prime movers

With the obvious availability of hydrocarbon gas as a fuel, and the requirement for a high power-to-weight ratio to keep structural scantlings to a minimum, gas turbines are the ideal prime movers for power requirements in excess of 1 MW. Below this value, reliability and other considerations (dealt with in Chapter 3) tend to make gas turbines less attractive to the system designer.

Owing to the complexity and relative bulk of gas turbine intake and exhaust systems, the designer is urged towards a small number of large machines. However, he is constrained by the need for continuity of supply, maintenance and the reliability of the selected generator set to an optimum number of around three machines.

A variety of voltages and frequencies may be generated, from the American derived 13.8 kV and 4.16 kV 60 Hz to the British 11 kV, 6.6 kV and 3.3 kV 50 Hz. Many ships operate at 60 Hz, including all NATO warships, and there is a definite benefit to be gained from the better efficiencies of pumps and fans running at the 20% higher speeds.

1.4   Key services or submain generators

On most platforms, smaller generators are provided to maintain platform power for services other than production. These are also normally gas turbine driven and can provide a useful blackstart capability, especially if this is not available for the main machines.

1.5   Medium-voltage distribution

The design of the distribution configuration at the platform topsides conceptual stage is very dependent on the type of oil field being operated and the economic and environmental constraints placed on the oil company at the time. The older platforms originally had few or no facilities for gas export or reinjection, and therefore the additional process modules installed when these facilities were required have their own dedicated high-voltage switchboards. This is also the case if the power requirement for such a heavy consumer as sea water injection is underestimated at the time of construction.

In general, however, it is better to concentrate switchrooms in one area of the platform in order to avoid complications with hazardous areas, ventilation etc., as discussed in Chapter 2.

4

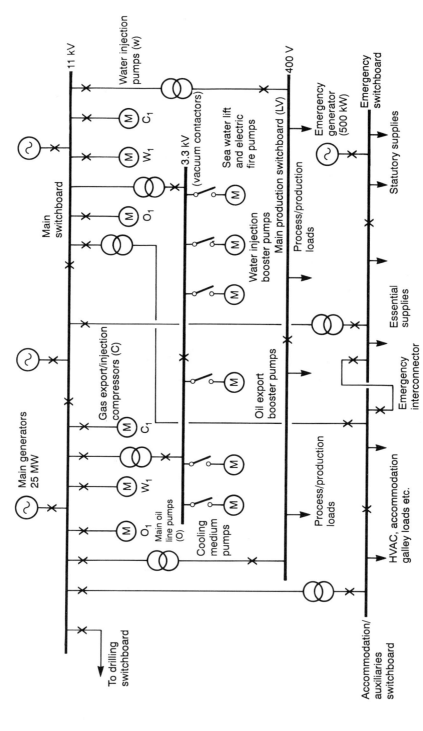

Figure 1.1 Single-line diagram of typical offshore electrical system

With such relatively high generation capacities and heavy power users within the limited confines of an offshore platform, calculated prospective fault currents are often close to or beyond the short-circuit capabilities of the MV switchgear designs available at the platform topsides design phase. Currently, fault ratings of 1000 MVA are available, and with careful study of generator decrement curves etc. it is usually possible to overcome the problem without resorting to costly and heavy reactors.

All the available types of MV switchgear are in use offshore. The use of bulk oil types, however, is questionable owing to the greater inherent fire risk.

Unlike land based switchboards, there has been found to be a significant risk of earth faults occurring on the busbars of offshore switchboards, and so some form of earth fault protection should be included for this.

The platform distribution at medium voltage normally consists of transformer feeders plus motor circuit breaker or contactor feeders for main oil line (MOL) pumps, sea water lift and water injection pumps, and gas export and reinjection compressors. Depending on process cooling requirements, cooling medium pumps may also be driven by medium-voltage motors.

Operating such large motors on an offshore structure (i.e. on the top of a high steel or concrete tower) can lead to peculiar forms of failure owing to the associated vibration and mechanical shock, almost unheard-of with machines securely concreted to the ground. This has led to offshore platform machines being fitted with more sophisticated condition monitoring than is usually found on similar machines onshore.

Another problem, which will be discussed in more detail in Chapter 9, is the transient effect on the output voltage and frequency of the platform generators with such large motors in the event of a motor fault, or for that matter during the normal large-motor switching operations. Computer simulation of the system must be carried out to ensure stability at such times, both at initial design and when any additional large motor is installed. Facilities such as fast load shedding and automatic load sharing may be installed to improve stability and also make the operator's task easier. This subject is discussed in Chapter 5.

1.6   Low-voltage distribution

Using conventional oil or resin filled transformers, power is fed to the low-voltage switchboards via flame retardant plastic insulated cables. Cabling topics are covered in Chapter 7.

Bus trunking is often used for incoming low-voltage supplies from transformers. Owing to competition for space, this is just as likely to be due to bending radius as to current rating limitations of cables, since bus ducting may have right angle bends.

The type of motor control centre switchboard used offshore would be very familiar to the onshore engineer. However, the configuration of the low-voltage distribution system, to ensure that alternative paths of supply are always available, is usually much more important offshore. This is because, although every effort is put into keeping it to a minimum, there is much more interdependence between systems offshore.

A few examples of small low-voltage supplies which are vital to the safe and continuous operation of the installation are as follows:

(a)   safe area pressurization fans;
(b)   hazardous area pressurization fans;
(c)   generator auxiliaries;
(d)   large-pump auxiliaries;
(e)   large-compressor auxiliaries;
(f)   galley and sanitation utilities for personnel accommodation;
(g)   uninterruptible power supply (UPS) systems for process control and fire and gas monitoring;
(h)   sea water ballast systems on tension leg platforms and semisubmersibles.

The topics of maintenance and availability are covered in Chapter 12.

1.7   Emergency or basic services switchboard

As a statutory requirement, every British Sector installation must have a small generator to provide enough power to maintain vital services such as communications, helideck and escape lighting, independently of any other installation utility or service. In the event of a cloud of gas enveloping the platform owing to a serious leak, even this may need to be shut down as a possible ignition source.

1.8   Fire pumps

Again as a statutory requirement, every installation must be provided with at least sufficient fire pumps with enough capacity to provide adequate water flow rates for fighting the most serious wellhead, pipeline riser or process fires. The numbers and capacities of these pumps have to take into account unavailability due to routine maintenance and failure. These pumps may be submersible electric, hydraulically powered, or directly shaft driven from a diesel engine.

Typically, one pump arrangement could have an electrically driven 100% capacity pump supplied from a dedicated diesel generator set which is directly cabled to the pump, i.e. with no intervening switching or isolating devices. This has the advantages of increased reliability due to fewer components and soft starting of the motor. This kind of pump runs up to operating speed with the generator, in the same manner as a diesel-electric railway locomotive would accelerate from start. The second 100% capacity pump could again be electric but supplied from the platform distribution system in the conventional way. The purpose of this arrangement is to avoid failure of both pumps owing to a common operational element, i.e. common mode failure. A third 100% capacity pump would be required to allow for maintenance downtime. Details on the electrical design of diesel-electric fire pump packages are given in Chapter 4.

1.9 Secure AC and DC power supplies

On any platform, there are a large number of systems which require supplies derived from batteries to minimize the risk of system outage due to supply failure.

The following is a typical platform inventory:

(a) fire and gas monitoring and protection;
(b) process instrumentation and control;
(c) emergency shutdown system;
(d) emergency auxiliaries for main generator prime movers;
(e) emergency auxiliaries for large compressors and pumps;
(f) navigational lanterns and fog warning system;
(g) emergency and escape lighting;
(h) tropospheric scatter link;
(i) line of sight links;
(j) carrier multiplexing and VFT equipment;
(k) telecommunications control and supervisory system;
(l) public address;
(m) general alarm system;
(n) platform PABX;
(o) marine radio telephones;
(p) aeronautical VHF (AM) radio;
(q) VHF (FM) marine radio;
(r) aeronautical non-directional beacon;
(s) company HF ISB and UHF (FM) private channel radios;
(t) telemetry system;
(u) satellite subsea well control systems.

The majority of these systems operate at a nominal voltage of 24 V DC and, although it is not necessary for each of the above systems to have separate battery and battery charger systems, the grouping criteria require more detailed discussion. These are covered in Chapter 5, as is the need to provide dual chargers and batteries for certain vital systems.

In addition to the above systems there are, of course, switchgear tripping/closing supplies and engine start batteries which are dedicated to the equipment they supply. In the case of engines which drive fire pumps, duplicate chargers and batteries are also required. This subject is discussed in Chapter 4.

1.10 Drilling supplies

A typical self-contained drilling rig supply system single-line diagram is shown in Figure 1.2.

The usual arrangement is for two or more diesel generators rated at around 1 MVA to feed a main switchboard, which also has provision for a supply from the platform generation and distribution system via transformers. This main switchboard then supplies a series of motor control centres, one or more of which contain a series of silicon controlled rectifier (SCR) direct current variable speed drive controllers. By using an arrange-

8

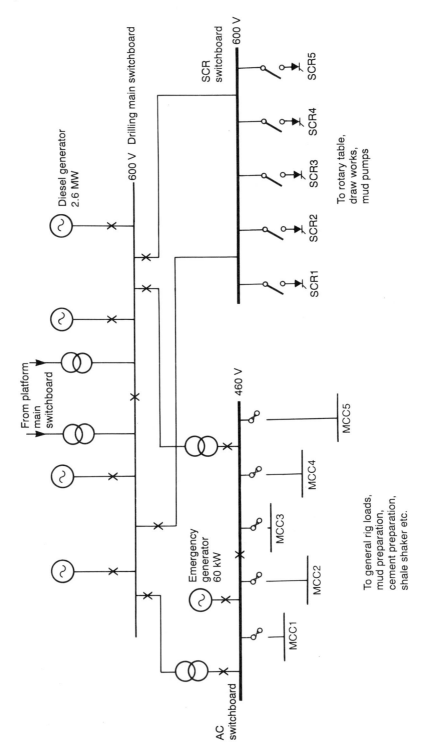

Figure 1.2 Single-line diagram of typical drilling electrical system

ment of DC contactors, these controllers may be assigned to various DC drive motors from the driller's console. As the SCR systems are phase angle controlled, a variable amount of harmonics is generated, depending on the kind of drilling operation being carried out. When the drilling rig is being fed from the platform supply, the harmonics may affect certain sensitive equipment, such as secure supply inverters.

Chapter 2

Electrical system general requirements

Some background information on oil and gas production is given in Appendix A. In this chapter it is the intention to discuss the more general criteria governing offshore electrical systems and equipment design.

2.1 Safety

The environment on an offshore installation is not inherently safe, owing to the heavily salt-laden atmosphere and the highly conducting nature of the structure and virtually all the equipment it contains.

It must not be possible for personnel to come into contact with live or moving parts either by accident or while performing their normal duties.

Protection against electrical shock relies on the safe design and installation of equipment, on training personnel to be aware of the dangers and to take the necessary precautions, and on the use of special safe supplies for most portable equipment.

An electric current of only a few milliamperes flowing through the human body can cause muscular contractions and, in some circumstances, will be fatal. The current may result in local burning or some involuntary reaction which in itself may lead to injury. Additionally, of course, varying degrees of pain will be experienced.

2.2 Isolated situation

Except in the case of one or two installations, the electrical system is totally isolated from any other means of electrical supply. The system must be designed and configured in such a way that it is never dependent on one small component or electrical connection to continue in operation. This point may sound rather obvious, but it is the author's experience that hidden vulnerabilities may be designed into systems which are both costly and disruptive in their first effects and in their eradication. The following examples of actual occurrences illustrate the point.

Example 2.1
A platform reinjection compressor is driven by a 500 kW 3.3 kV motor having a lube oil system pressurized by auxiliary lube oil pumps driven by low-voltage motors. Both the lube oil pump motors and the control supply for the 3.3 kV latched contactors are supplied from the same low-voltage switchboard. A spurious gas alarm occurs in the vicinity of the low-voltage switchroom and the LV switchboard incoming circuit breakers are opened by the emergency shutdown system. The lube oil pumps then stop but the compressor control system is unable to open the main drive motor contactors and the motor runs to destruction.

Example 2.2
A platform has two low-voltage switchboards dedicated to providing the safe and hazardous ventilation necessary for continued safe operation of the platform. Depressurization of any module would lead to a process shutdown. The particular platform is a pumping station for oil from other platforms, including those of other companies, and therefore considerable oil revenue is at stake if the platform is shut down. Unfortunately each switchboard is fed by a single incomer, and the ventilation fan motor starters are distributed so that the majority of supply fans are on one switchboard and the majority of extract fans on the other. This arrangement resulted in the export of oil from a number of large North Sea installations being dependent on the continuous operation of two small low-voltage switchboards.

The subject of reliability is dealt with in greater detail in Chapter 12.

2.3   Environment

This topic is covered in greater detail in Chapter 8, and is exhaustively covered in all the relevant standards, recommendations and codes of practice (given in the Bibliography). However, it is important to be very clear as to the fundamental reasoning behind all the regulations governing electrical installation offshore. Because both the safety and the cost of an installation are highly sensitive to equipment selection, it is also important to have a clear understanding of the reasons behind the classification of hazardous areas and of the different methods employed by equipment manufacturers to make their equipment suitable for particular environments.

   Where this is practicable, electrical equipment is best installed in an environmentally controlled room which is located in an area unclassified with respect to hydrocarbon gas ignition risk, is effectively sealed from the outside atmosphere, and is provided with a recirculating air conditioning system. Of course, this optimum scheme cannot be considered for equipment which:

(a)   has to be located outside (such as navigational aids);
(b)   has to be located under or near water (such as sea water lift pump motors); or

(c)   is associated with some other equipment which may occasionally or does normally leak hydrocarbon gases (such as gas compressor drive motors).

Often the equipment installed has to safely cater for a combination of all three situations, and may also be required to operate at elevated pressures and temperatures.

2.4   Water hazards

Hazards due to water coming into contact with electrical equipment are similar to those experienced on ships, but can be more catastrophic since more power is generated at higher voltages with greater prospective fault ratings.

Water may leak from large-bore water carrying pipes routed over switchboards or generators. The following are two examples seen by the author where such pipes were routed over switchboards.

Example 2.3
A fire water main was routed over a 4 MW gas turbine generator. The fire main was flanged and valved directly over the alternator. When the valve was serviced while the generator was running, the pipe fitter inadvertently drained an isolated section of fire main over the alternator. The generator promptly shut down, owing to operation of the differential protection, with a stator winding fault.

Example 2.4
In an accommodation module, a sewage pipe from an upper floor was routed directly above a low-voltage switchboard and along almost its entire length. Although some minor leaking had taken place, it was fortunate that the only problem for the electrical maintenance staff was one of hygiene.

If routing of water carrying pipes over switchgear is unavoidable, there should be no flanges in the section of pipe over the switchboard.

2.5   Hydrocarbon hazards

In the planning of platform superstructures, designers try to arrange to segregate the wellhead and process areas from the accommodation and other normally manned areas to the greatest possible extent. This involves not only horizontal and vertical segregation but also segregation of all piped or ducted services such as ventilation ducting and drains.

Following the Piper Alpha disaster, it is likely that the whole philosophy regarding the segregation of accommodation areas on offshore platforms will be rethought. As is common knowledge, 165 men lost their lives either as a result of the initial explosions, dense smoke and fire, or following the ensuing riser fires which led to the loss of structural integrity and the falling of the accommodation modules into the sea. The recently published Cullen

Report gives over 100 recommendations, covering all aspects of offshore installation design, construction, operation and safety. In one of (in my view) the most important recommendations, Lord Cullen states that the operator should be required by regulation to submit to the regulatory body a safety case in respect of each of its installations. It is important to consider the safety aspects of each installation uniquely so as to meet objectives, rather than to impose fixed solutions which may or may not work on a particular installation.

Whatever further means of ensuring the survival of the particular installation and its personnel are considered in the safety case, it is certain to influence the design of the electrical system and equipment, particularly in minimizing the risk of electrical ignition sources and in the provision of emergency secure electrical supplies completely independent of normal platform supplies.

An important means of minimizing the risk of ignition sources are the hazardous area boundary drawings produced during the platform process design stage, which represent the situation during normal operating conditions (see Chapter 8). However, it is also necessary to consider the situation during a major outbreak of fire or after a serious gas leak – the so-called 'post-red' situation.

There are three systems which normally monitor and control the extent of oil and gas leaks and hence the safety of the platform:

(a)   the fire and gas monitoring system;
(b)   the emergency shutdown system;
(c)   safe and hazardous area ventilation systems.

On floating installations, the ballast control system could also be included as a fourth in certain circumstances.

All these systems will have some bearing on the design of the platform electrical system, either because they may include the facility to shut down all or part of the electrical system, or because a secure (or at least a more reliable) electrical supply is needed to operate them.

2.5.1 Fire and gas monitoring

Every installation must have, as a statutory requirement, a designated control point located in a non-hazardous area, capable of overall management of the installation and manned continuously.

All pertinent information from the production processes, drilling, utilities and fire fighting systems need to be monitored at this control point, and emergency controls associated with these systems have to be available there to enable sufficiently effective control to be exercised in all operational or emergency conditions.

On normally manned installations, the control point needs to be located in or adjacent to the accommodation area, and may be in or adjacent to the offshore installation manager's (OIM's) office and radio room. This control point also requires public address facilities to be close at hand. As a temporary safe refuge (TSR) it would normally be the last area to be vacated in an emergency, and the room in which it is contained would be H120 fire rated, with a dedicated ventilation system (see Chapter 8).

In the larger platforms this is a limited repeat of a far more sophisticated monitoring system located in or near the process control room.

On normally unmanned platforms the basic control point may be located on an adjacent normally manned installation or even at a control centre onshore.

2.5.2 Emergency shutdown (ESD) system

As in the case of a nuclear reactor or similar complex system, the continuing safe condition of the platform cannot be left solely to the human operators, since they would not always have sufficient time to investigate each abnormality and respond with the appropriate sequence of corrective actions in every case. Because of this, it is necessary to provide a system which either initiates the correct sequence of actions itself, or provides a series of simple options (levels of shutdown) that the operator may take when a particular event occurs.

Every installation has startup and shutdown systems of varying sophistication, which attempt to provide the greatest possible safety for personnel and equipment. These systems are interrelated, with the process control system being subordinate to the emergency shutdown system. The various levels of shutdown and their effects on the electrical system are discussed in Chapter 5.

2.5.3 Safe and hazardous area ventilation

Most of the heating, ventilating and air conditioning (HVAC) systems must run continuously during normal platform operation to ensure that:

1. Acceptable working environments are maintained in process modules containing equipment or pipework which may leak hydrocarbon gas.
2. Comfortable environmental conditions are maintained within the accommodation modules and normally manned non-hazardous areas, and an acceptable working environment is provided in normally unmanned modules.
3. Positive pressurization with respect to adjacent hazardous areas or the outside atmosphere is maintained in non-hazardous modules or rooms.
4. Potentially hazardous concentrations of explosive gas mixtures are diluted in, or removed from, hazardous area modules.
5. Individual areas are sealed from ventilation and the associated fans are shut down in accordance with the logic of the emergency shutdown system, when fires occur or dangerous concentrations of gas are detected.
6. Uncontaminated combustion, purging and normal ventilation air is available to prime movers.
7. Uncontaminated air supplies are available to personnel, and emergency generator and fire pump prime movers and other essential service equipment are provided with combustion and ventilation air in times of emergency.

Ventilation systems, especially those associated with switchrooms and generator rooms, are discussed in more detail in Chapter 8.

2.5.4 Combined action

The three systems – fire and gas, emergency shutdown, and HVAC – are interconnected and are often required to work in concert. An example of this would be if a fire occurred in a particular switchroom. The fire would be detected by smoke or heat detectors, and the central fire and gas system monitoring the room would initiate the following actions:

(a) signal the ventilation system to seal the room by the closure of ventilation fire dampers and switch off associated fans;
(b) sound an alarm in the switchroom to warn personnel that escape is necessary and that a fire extinguishant is to be released;
(c) depending on the system logic, signal the emergency shutdown system to isolate the switchboards in the switchroom by opening the appropriate feeder circuit breakers, or even shut down all main generators if the switchroom in question contains the main switchboard;
(d) release the fire extinguishant (CO_2 or halon gas) into the switchroom after a suitable time delay to allow for personnel to escape.

2.6   Distribution configuration

Figures 2.1 to 2.3 show various attempts to obtain optimum availability from the platform electrical power system within the limitations of weight and the conditions imposed by the situation.

Figure 2.1 shows the system arrangement on an earlier installation, consisting of a main 6.6 kV switchboard with two 15 MW gas turbine generators and a smaller 6.6 kV switchboard with two 2.5 MW gas turbines. In this configuration, the smaller utilities switchboard provides supplies for all cooling sea water pumps and also feeds all low-voltage distribution transformers except those in a water injection module which was added to the platform at a later date.

Should a fault occur affecting the whole of the utilities switchboard, for example a fault in the bus sections or sectioning switch leading to busbar damage, then only the emergency generator is left available. This is because there is insufficient spare capacity in the emergency generator to run the main generator auxiliaries, and in this particular case the main generators require sea water for cooling, normally available from the sea water lift pumps powered from the utilities switchboard.

The probability of water ingress, the most likely cause of such a failure, would be reduced if the switchroom housing the utilities switchboard was surrounded by other rooms or modules.

Figure 2.2 shows a scheme where the main switchboard consists of a single-section switchboard fed by two 24 MW gas turbine generators, each capable of taking the entire platform load including drilling. This medium-voltage switchboard, as well as directly supplying the platform large-motor

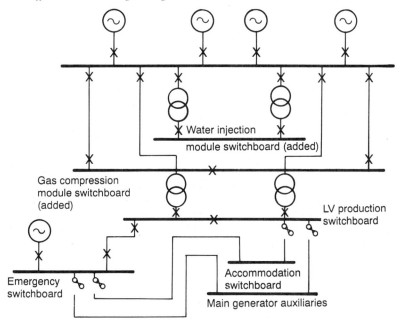

Figure 2.1 Single-line diagram of early installation with added utilities switchboard

drives such as MOL pumps and gas compressors, also feeds the majority of lower-voltage switchboards via transformers.

There are two points at which connections are made to the drilling electrical system. One is a supply at medium voltage, and the other is an alternative supply to the emergency switchboard at 440 V. The drilling system is also supplied by two dedicated 1.5 MW diesel generators, but these are much too small to sustain platform production of any sort.

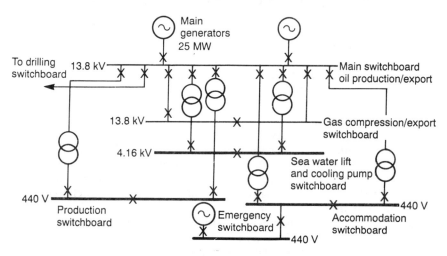

Figure 2.2 Single-line diagram of system with two platform load capacity generators

It can be seen that production is dependent on the continued health of the main switchboard, a single fault on which causes the majority of the installation to be blacked out. The insertion of a bus section switch in this particular case would do little to improve the system integrity, as the reliability of the switch is likely to be less than that of the busbars.

A configuration similar to that shown in Figure 2.3 for the main switchboard is now the most favoured for offshore use, since, assuming supplies are maintained to generator auxiliaries, it requires several failures to occur before production (and hence revenue) is affected.

2.7   Modular construction

Provided lifting and transport facilities of sufficient capacity are available at an economically viable cost, it is invariably better to build a complete module containing the generators, switchgear and transformers, completely fitted out, tested and commissioned at a suitable fabrication yard, than it is to carry out any of the construction on the platform.

Apart from fuel, cooling air and combustion air, it is preferable to make the electric power module totally independent of the rest of the installation. This also means that it has integral engine starting facilities as well as engine auxiliaries which do not require external low-voltage electrical supplies. This is not always possible owing to the weight of the extra transformers required. The need for sea water cooling for alternator heat exchangers may be unavoidable, owing to the bulk of air cooled units. The optimum independence of the module has the added advantages of:

(a)   minimal hookup requirements during offshore installation;
(b)   minimal service requirements during test and commissioning at the module fabrication yard.

2.8   Subsea cable versus onboard generation

The cost of laying subsea power cable is at present about £2 million to £5 million per cable kilometre. This cost includes that involved in mobilizing a suitable cable laying vessel. At first sight this may appear prohibitively expensive, but it is worth investigation if most of the following conditions are met:

1.   The source of supply is conveniently at hand, i.e. on the mainland or another installation. Subsea cable lengths greater than 50 km are unlikely to be viable.
2.   The cable route is not through a busy shipping lane or anchorage where there is a high risk of anchor damage.
3.   The cable route does not pass through a popular fishing ground where there is the consequent risk of cables being dredged up by trawl nets or gear.
4.   The fuel gas supply on the installation to be supplied by the cable is unreliable, exhausted or of too small a capacity for the required prime

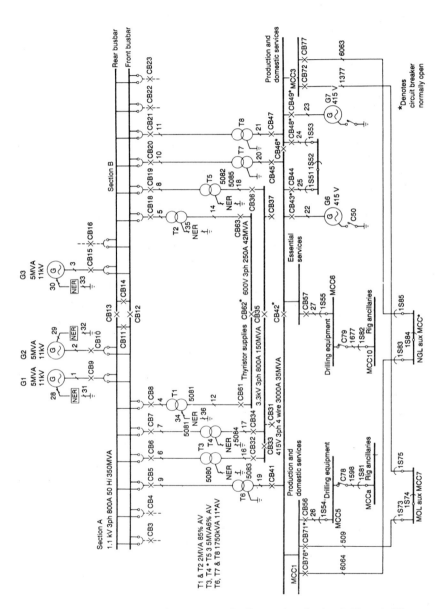

Figure 2.3 System with three main generators feeding sectionalized switchboard. (Courtesy BP Exploration)

mover. Of course, the converse must be true of the supplying installation.

5. The transmission route is through reasonably sheltered waters where it is probable that cable repairs could be carried out throughout the year. If this is not the case then the expense of laying duplicate cables over separate routes becomes more attractive, since in winter it could be three months or more before a suitable weather window becomes available to repair a cable fault.

6. The sea floor is suitable for burial of the cable as a means of protection. Drifting sands, solid rock or strong tidal currents would militate against this.

In listing the above conditions, it has been assumed that the importing of diesel fuel for normal operation is not a viable option owing to the high cost involved. However, this may depend on commercial considerations such as the revenue value of the gas to the installation operators. For example, British Gas's Morecambe Bay complex has a mixture of subsea cable, gas turbine and diesel generator powered platforms.

If it becomes necessary to supplement the available power on a particular platform, then the additional weight of supplementary generators may be too great for the platform to bear without very costly modifications. Even if weight is not a problem, it is not always possible to find a sufficiently spacious location on the installation.

Shore supplies may be of the wrong frequency for use on the particular installation, and it may be necessary to install a motor-generator set. This has the additional advantage of improving the motor starting capability of the supply, as the generator impedance will be much lower than that of a series of transformers and long subsea cables.

The transmission voltage required will vary depending on the length of the subsea cable, but is likely to be either 11 or 33 kV. The weight and space taken up by the transmission transformers and the associated extra switchgear need to be taken into consideration whenever subsea cable options are proposed.

If there is a group of several small installations separated only by a few kilometres of water, it may be economic to supply all their main power requirements from one central platform. This is more likely to be the case if centralizing the main generation allows gas turbines of 1 MW or more to be considered (see the discussion of prime mover selection in Chapter 3).

Finally, it is advisable to carry out some form of reliability analysis in order to numerically rank the reliabilities of various supply and generation schemes before making a final decision. Reliability topics are discussed in Chapter 12.

Chapter 3

Prime mover and generator selection

In this chapter, the criteria for the selection of prime movers and generators for various applications are addressed. Before the particular criteria for selecting an engine are outlined, the prime mover types are described.

3.1   Gas turbines

Although there are some smaller machines installed, it is not common practice to install gas turbines of less than 1 MW offshore. This is for two reasons. First, gas turbine reliability generally tends to improve at around this size for machines in continuous operation. Secondly, the bulk of the intake and exhaust ducting involved in order to handle the large volume of air required, to reduce the noise to acceptable levels and to protect the engine from the marine environment, tends to make diesel or gas ignition engine prime movers more attractive up to this size.

There are two forms of gas turbine in use:

Aero engine derived   This consists of a modified aeronautical jet engine known as the gas generator which exhausts into a separate power turbine. This combination often produces a unit with very good power-to-weight ratio, as the gas generator is lighter than the integral unit on the equivalent industrial machine. However, the unit may require better protection from the environment and in some cases shorter intervals between servicing. Examples of this type of machine are Rolls-Royce Avon and RB211 based sets. Figure 3.1 illustrates a typical example.

Industrial   These are purpose-built engines, which incorporate the gas generator and power turbine in a single design. The older machines tended to be less fuel efficient than equivalent aero engine derived types, but they have a good reputation for reliability and for the toleration of fuel supply or load abnormalities. Examples of this type of machine are the General Electric and John Brown Frame 5 and the Ruston TB series. Figure 3.2 illustrates a typical example.

(a)

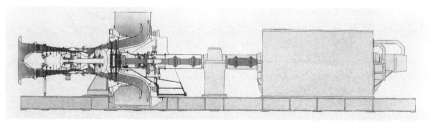

(b)

Figure 3.1 Aero engine derivative generator set. (a) Rolls Royce industrial Avon powered compression set being installed on the Brunei Shell Petroleum Company platform; (b) sectional drawing of set shown in (a). (Courtesy of Crest Communications Ltd.)

22

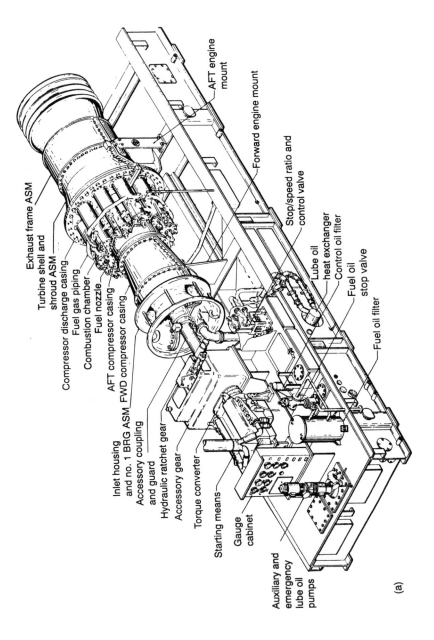

Exhaust frame ASM
Turbine shell and
shroud ASM
Compressor discharge casing
Fuel gas piping
Combustion chamber
Fuel nozzle
AFT compressor casing
FWD compressor casing
Inlet housing
and no. 1 BRG ASM
Accessory coupling
and guard
Hydraulic ratchet gear
Accessory gear
Torque converter
Starting means
Gauge
cabinet
Auxiliary and
emergency
lube oil
pumps

AFT engine
mount
Forward engine mount
Stop/speed ratio and
control valve
Lube oil
heat exchanger
Control oil filter
Fuel oil
stop valve
Fuel oil filter

(a)

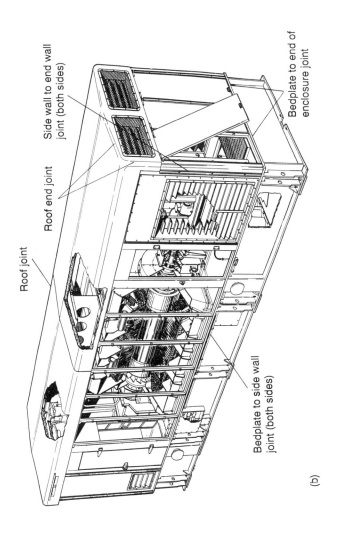

Roof joint

Roof end joint

Side wall to end wall joint (both sides)

Bedplate to end of enclosure joint

Bedplate to side wall joint (both sides)

(b)

Figure 3.2 Industrial engine based generator set; (a) engine; (b) generator. ((a) Courtesy of John Brown Engineering (to GE (USA) design); (b) Courtesy of Hawker Siddeley Electrical Machines Ltd.)

3.2   Gas turbine application

The following can be applied to both forms of gas turbine, and is designed to assist the electrical engineer in the selection and application of gas turbines as prime movers. It should be noted that there are many other considerations, beside those mentioned below, involved when installing a turbine; as they are of no direct concern to electrical engineers, they are beyond the scope of this book. However, some reference is made to fire fighting facilities in generator rooms in Chapter 5.

3.2.1 Fuel gas supply disturbances

In most offshore situations (with the exception of storage and pumping stations) the gas is being produced via the process plant from a production well. Changes in well gas content, calorific value, pressure etc. can have significant effects on engine power output, which may in turn affect the production process and the fuel supply. Slugs of condensate may also appear in the fuel gas supply, having a similar effect. Although these unpredictable phenomena are avoided as much as possible by good process design, it is worth considering some means of catering for them such as one or more of the following remedies:

1.   A small separator or knockout pot may be located close to the engine intake in order to remove condensate. If the supply pipe is long, it will be necessary to provide this anyway to remove condensate which has condensed in the pipe.
2.   The engine combustion system may be modified for dual-fuel operation. When set up properly, this system can automatically transfer from gas to diesel combustion during fuel gas disturbances with negligible effect on engine power output.
3.   If there is a problem with fluctuations in the calorific value of the gas to the extent that the electrical load cannot be met on a downward fluctuation, it may be necessary to install a fast acting automatic load shedding system as described in Chapter 5.

3.2.2 Turbine temperature limits

Gas turbines depend for their cooling on the vast quantity of air which passes through them. A basic design limitation on the operation of any gas turbine is the operating temperature of the power turbine blades. Disastrous changes in their mechanical properties will occur should they get too hot. By monitoring the exhaust gas temperature using a sophisticated control system, such as the GE Speedtronic or the Ruston Rustronic governor, it is possible to bring the engines up to power automatically and to continuously control the supply of fuel in such a way that the maximum exhaust temperature is never exceeded.

As the cooling effect of the incoming air is proportional to its density and temperature, altitude and ambient air temperature have a very significant effect on available output power. Altitude, of course, is fixed, but if the engine is running close to its rated power output then a small increase in ambient air temperature may cause the engine control system to limit the fuel supply in order to prevent the maximum exhaust temperature being exceeded. As it is unlikely that the generator load would have been reduced at the same time, the effect of this fuel supply reduction is for the generator set to begin slowing down until an underfrequency trip occurs. Even in the North Sea, warm weather conditions can sometimes reduce turbine power output capability below the required rating. The problem is often exacerbated by the poor location of combustion air intakes and exhausts.

The following must be considered:

1. Intakes should be located as far as possible from any engine exhaust, including that of the engine under consideration.
2. The effects of any process flares must be taken into consideration, for both hot gas and radiation.
3. The effect of wind on all the various platform exhausts must also be considered. Although the prevailing wind is the most important consideration, the turbine must be able to develop sufficient power in any wind condition, and with any combination of other engines installed on the platform, working at their normal outputs.
4. If the exhaust plume from another engine causes the engine in question to govern down, this may be overcome by the use of an extra ducting or a water curtain installed around the exhaust of the other engine.

Air flow through the engine and hence cooling may be improved by cleaning the compressor section of the engine after a few months of operation. The improvement in output power after routine cleaning is usually significant and can be as much as 10% of its rated power. The selectors of the prime mover must take this into account in their rating calculations.

If it is necessary to install a load shedding system because of engine power limitations or increasing electrical demand, it can be seen from the above that the system should take turbine exhaust temperature into account. If the load shedding system tripping level is based purely on monitoring electrical load for a fixed worst case value, production operations may be unnecessarily curtailed. In colder weather conditions as much as 15% of extra engine capacity would still be available. This may amount to several megawatts with a 25 MW generator set. The converse is also possible, when exceptionally warm weather conditions may reduce engine capacity to below the load shedding system tripping point, causing the generator set to trip on underfrequency.

3.2.3 Gas ingestion from leaks

A large gas turbine generator set may take in the region of 30 seconds, from the initiation of an emergency stop signal, to slow down to a speed at which acceleration is no longer viable without cranking. During this period of deceleration, it is vulnerable to ingestion of gas from serious process gas leaks on the platform. Such ingestion of gas in sufficient quantity, bearing in mind that all electrical loads would probably have been removed by the

ESD systems, may cause the machine to reaccelerate and overspeed to destruction.

The risk of this occurring must be minimized. First, there should be careful siting of gas detectors to ensure that serious gas leaks are detected as soon as possible. A 2-out-of-N voting system should be used to minimize spurious generator trips due to faulty detectors. Secondly, governor response to an emergency engine stop signal should be as fast as possible within the metallurgical limitations imposed by turbine blade cooling rates etc.

3.3   Reciprocating engines

3.3.1 Diesel engines

By far the most common engines for small to medium power requirements offshore, as well as prime movers for alternators, the diesel engine will be found directly driving anything from cranes to fire pumps. Diesel engines larger than 1 MW are rarely found, however, for two reasons. First, weight and vibration problems may be encountered in the platform structure. Secondly, if the engine is driving an alternator which provides normal production, i.e. non-essential supplies, and gas of sufficient quantity and quality is available during production, then the importing of large quantities of relatively expensive diesel oil is likely to be commercially unattractive. It would also require heavy storage tanks located in a site that would not constitute a fire hazard, which could well be a problem in the case of a steel structure.

It is possible to reduce the quantity of diesel oil consumed by burning a mixture of diesel oil and fuel gas. The ratio of gas to diesel is limited to approximately 85%, the limitation being due to the amount of diesel required to be injected in order to preserve the compression ignition action of the engine. This limitation has the benefit that a governed diesel engine is unlikely to overspeed due to the ingestion of gas from a process gas leak.

3.3.2 Gas ignition engines

If it is required that the engine should run on fuel gas only, then an ignition system will be required similar to that found on petrol engines. It is normal, however, in order to improve reliability, to use a low-tension distribution system with individual coils mounted over the spark plugs on each cylinder, known as a shielded system.

Reliability is improved by the following:

1. There is reduced line loss in the coil secondary circuit, as it is very short. Plugs and coils screw together to form an integrated unit.
2. The probability of earth faults occurring in the secondary circuit is substantially reduced, as there are no high-tension cables routed close together.
3. Common routeing of high-tension cables can also lead to transformer effects, which result in ignition voltages appearing in the wrong cable at the wrong time, engine misfiring and loss of power.

3.3.2.1 Fuel gas pressure
As with a conventional petrol engine, the fuel-to-air ratio is controlled by means of a carburettor, and the fuel gas pressure of a few psi has to be accurately controlled.

3.3.2.2 Fuel gas temperature and combustion knock
Natural gas is predominantly methane, which has a high octane rating and is therefore quite resistant to combustion knock (pinking). However, factors which increase charge temperature will also increase the likelihood of knock occurring. Such factors are:

(a) compression ratio;
(b) fuel gas temperature;
(c) high ambient air temperature;
(d) charge air temperature output from the turbocharger intercooler (where this is fitted).

There are other factors associated with the design of the engine which would affect how prone the engine is to combustion knock. It is important to ensure that a manufacturer's warranty is provided, which states that the engine will run at the rated power output without any shortening of operational life, for the fuel and conditions expected on the platform. It would be prudent to provide the manufacturer with a recent platform fuel gas analysis if this is available.

3.3.2.3 Spark duration and voltage
As with most spark ignition engines, it is important that the sparks are of sufficient voltage and duration to ensure good combustion. With ignition engine generator sets, poor sparking and the consequent poor combustion will give erratic speed control. The effect is noticeably different from the regular hunting associated with governor control problems in that it appears as a random speed wandering over a few tens of rpm. The problem will tend to be worse at lower loadings and can make it difficult to obtain satisfactory operation from autosynchronizing equipment.

3.3.2.4 Fuel gas disturbances
Fuel gas is tapped from the platform process separators at often very high temperatures and pressures, and these must be reduced to values within the operating envelope of the prime movers being supplied. Instrumentation will be required to ensure that gas at the wrong temperatures and pressures cannot reach the engine. This would mean the automatic operation of isolation valves and the shutdown of the generator set.

Occasionally, fluctuations in the quantity of condensate may pass through the process system and reach the engines. Slugs of condensate may, as discussed earlier, cause a speed fluctuation in a gas turbine, but with a gas ignition engine an 'incompressible' liquid present in one or more cylinders will almost certainly result in serious damage. Therefore it is vital to provide near the engine a small separator or knockout pot designed to remove both slugs of condensate and any liquid which has condensed on the walls of the fuel gas supply pipe. A high liquid level in the knockout pot should be arranged to shut down the engine.

3.4   Load profiles

The following topics need to be taken into consideration when selecting the number, type and rating of generator sets on a particular installation.

3.4.1 Projected demand

Over the life of the platform, the generation requirements may double or even triple as each new operational phase is reached. A typical platform demand profile is as follows:

Operational phase	Power demand
Drilling	3 MW
Oil export	12 MW
Gas compression/export	20 MW
Artificial lift	25 MW

3.4.2 Variability of demand over 24 hours

On a large oil production platform, the larger power users such as water injection pumps, main oil line pumps and gas compressors constitute the majority of the electrical demand. This will remain constant over 24 hours unless some planned change of plant is necessary or a breakdown occurs.

If drilling activities are powered from the main platform system, some quite large but transient demands, possibly of the order of a few megawatts, can be expected from the rotary table or draw works when difficult mineral formations are experienced.

On small installations where gas is exported without the need for compression, and where there are no round-the-clock maintenance shifts, a distinct profile will be created by the use of galley equipment and electric water heaters etc. in the accommodation areas.

3.4.3 Low loading problems

No internal combustion engine will run very efficiently at loads much below 50% of its rated full power output. This is particularly a problem with reciprocating engines since, below about 30% of full power, combustion products such as soot and gum will begin to collect inside the engine in sufficient amounts to substantially reduce the power available from the engine when the load demands it. Turbochargers are particularly susceptible and will be stopped by deposits after only a few hours running at low loads. If the load profile dips dangerously low for an hour or so, but there is then a period where the load is substantially higher, i.e. greater than 50% of full power, it is likely that the increased combustion will clear the engine of the buildup in these deposits.

As with a fuel gas analysis, it is important that a cyclic load profile is presented to the engine manufacturer. A statement should be obtained from the manufacturer to the effect that the engine will continue to run without deterioration with the load profile submitted.

Should such guarantees not be forthcoming, it will be necessary to reconsider some of the other engine options and/or whether the most suitable number and rating of generator sets have been selected.

3.5 Choice of fuel

The choice of fuel is usually governed by the following:

1. The quality and quantity of gas available from the field being exploited. Where the reservoir produces a preponderance of oil and there is a likelihood that there will not always be sufficient gas for fuel, then some other fuel will be required, either continuously or as a standby.
2. Availability of well gas at the times when the engine is required to run. If the engine is the prime mover for an emergency generator, it will need to run when the production process is shut down and no gas supply is available.
3. Logistical costs associated with the transport of diesel oil to the platform concerned. The use of diesel oil requires that sufficient quantities of diesel can be stored on or in the structure to allow for periods of bad weather when refuelling is not possible. As this fuel will constitute a considerable fire risk, the storage location will need to be carefully considered.
4. Comparative costs of connection to a suitable power generation complex. This may be a nearby platform with spare capacity or an onshore supply system. The costs may also need to include those for the purchase and installation of a suitable motor-generator set to cater for platform supply frequency and/or to effectively reduce the supply system impedance.

3.6 Main generation

When all the limitations such as weight and dimensions imposed by the platform structure have been established, the supply and operating constraints can be tackled.

3.6.1 Number required

From the reliability point of view, three generators, each rated for the full platform load, are the optimum. This allows for one generator running, one on standby and one undergoing routine maintenance. If space is in particularly short supply, it may be necessary to dispense with the third machine. If point loading on the platform structure is a problem, it may be necessary to use a greater number of smaller-rated machines.

3.6.2 Size

The types of prime mover available have already been discussed in this chapter. One of the benefits of using large machines, each capable of supplying the whole system load, is that of improved motor starting capability and greater stability during power system disturbances.

However, some care will have to be exercised to ensure that alternator subtransient reactances on such large machines are high enough that switchgear of sufficient fault capability can be obtained without going outside the normal manufactured ranges. Few manufacturers produce standard switchgear at voltages in the 11 to 15 kV region above 1000 MVA. Cables capable of withstanding the associated prospective fault currents would have to be sized very much over their current ratings and would be expensive, heavy and more difficult to install. Thus 1000 MVA is considered a practical top limit for system fault levels at present.

3.6.3 Location

Generator modules should be located in an area classified as safe when the release of flammable gases is considered. This is necessary to reduce the risk that flammable gas might be drawn into the engine enclosure and be ignited on hot parts of the engine. If a small quantity of gas is drawn into the engine intake, this should not cause a significant increase in engine speed as the governor should correct for the presence of this extra 'fuel'. Large concentrations of gas, however, may cause overspeed in gas turbines. In all engines they may interfere with combustion and, if no precautions are taken, may be ignited in the engine air intakes, leading to fires and/or explosions.

In general, the following precautions must be taken with gas fuelled engine enclosures:

1. Any part of the engine, including such ancillaries as turbochargers, exhaust systems and anything else in the enclosure which may have a surface temperature in excess of 80% of the ignition temperature of the actual gas/air mixture (200°C in the case of North Sea gas), must not under any circumstances be allowed to come in contact with such an explosive gas/air mixture. This can be avoided by:
    (a) providing sufficient ventilation to prevent gas accumulating;
    (b) enveloping the hot areas in a water cooling jacket (note that it is often impractical to do this with a turbocharger)
    (c) ensuring that the enclosure is always positively pressurized (with air free from gas), so that gas cannot be drawn in from outside the enclosure;
    (d) keeping potential gas leak sources in the enclosure (such as fuel gas pipe flanges) to an absolute minimum.
2. A 'block and bleed system' should be included in the fuel gas supply system so that when the engine is stopped, the entire length of fuel gas supply pipe within the enclosure is blocked by isolating valves at both ends and vented safely to atmosphere.
3. Electrical instrumentation and controls associated with the fuel gas pipework, such as pressure and temperature transmitters, solenoid

valves and throttle actuators, should all be suitable for safe use in areas where explosive gas mixtures may be present. With spark ignition engines, the shielded ignition system discussed earlier in this chapter should be used. This is because the high ignition voltages are only present within the engine with this type of system.

4.   In reciprocating engines, various extra precautions need to be taken as follows:

(a)   Drive belts must be of the anti-static, fire resistant type.

(b)   Cooling fan blades must be of a type which cannot cause friction-sparks if they come in contact with adjacent parts.

(c)   Exhaust systems must be fitted with flame-traps and spark arresters.

(d)   A flame trap may be required in the combustion air intake in order to protect against a flashback through the induction system (i.e. backfiring). This will be required even if the intake is in a safe area, if the engine may be run in emergency conditions.

(e)   All diesel engines should be fitted with a 'Chalwyn' or similar air induction valve, to prevent overspeeding if flammable gases are drawn in with the combustion air. This is particularly important if the engine runs on a fixed fuel-rack setting, i.e. is not fully governed.

(f)   If the crank case volume is greater than 0.5 m^3 relief valves must be fitted to the crankcase to prevent damage or external ignition due to crankcase explosions. The relief device must be provided with its own spark arrester/flame trap system.

(g)   Care must be taken to ensure that any special design features that could cause external ignition are adequately catered for. For example, turbochargers must be water-cooled; decompression ports, if absolutely neccessary, should be treated in the same way as exhaust systems.

(h)   Engine governors, and the fuel injection pumps of diesel engines, should be designed so as to make reverse running of the engine impossible.

For further information, please refer to Publication MEC-1 of the Engineering Equipment and Material Users Association (formerly OCMA) entitled *Recommendations for the Protection of Diesel Engines Operating in Hazardous Areas*.

The subject of hazardous areas is discussed in greater detail in Chapter 8.

3.7   Cooling systems

Although this subject and most of the following ones in this chapter are in the realms of mechanical and other engineering disciplines, the electrical engineer needs to keep a weather eye on the proceedings or risk being rudely awakened when problems such as insufficient cooling arise during commissioning of the generator module.

At some stage during purchase and manufacture of the generator set, the manufacturer will provide heat balance figures. Typical heat balance figures are as shown in Table 3.1.

Table 3.1 Heat balance values

	Energy gain (kW)	Heat loss (kW)	Electrical output (kW)
Energy provided by fuel	1500		
Heat lost to engine water jacket		450	
Heat radiated from engine		50	
Mechanical output from engine:			
Losses from alternator (eff. 90%)		50	450
Heat lost to turbo intercooler		50	
Heat lost to engine exhaust		450	
Totals	1500	1050	450

As with an accountant balancing his books, the engineer must account for all the waste heat from the engine in the design of cooling and ventilation systems. The ratings of the engine and alternator are based on designed operating temperature bands; if these are exceeded when the engine is running at its rated power output because of poor cooling and ventilation, the generator manufacturer will have to derate his equipment accordingly. The heat balance example in Table 3.1 is for a reciprocating engine, but the same principle may be applied to turbines.

In generator modules where all ventilation is provided by the engine radiator fan, an allowance must be made for the temperature rise caused by heat dissipated within the module before the air flow reaches the engine radiator. Wind speed and direction will also affect the air flow through the module and, when the wind is strong and blowing directly against the fan, may stall the air flow completely, causing a rapid temperature trip. If the radiator fan is electric, windmilling of the fan should be prevented when the engine is not running, otherwise when the engine is started the fan motor may trip on overload owing to the excessive acceleration time from some negative to full forward speed. If power is available from another source after the generator has stopped, it is advisable to have another smaller fan running to prevent a buildup of heat in the compartment while the generator set is cooling down. Without this, the temperature in the compartment may exceed maximum allowable values for electrical equipment or insulation.

3.8   Lube oil systems

On large gas turbine generator sets, lubrication is accomplished by a forced feed lube oil system, complete with tank, pumps, coolers, filters and valves. Lubricating oil is circulated to the main bearings, flexible couplings and gearboxes. A portion of the oil may be diverted to function as hydraulic oil for operation of guide vanes etc. within the turbine. A typical arrangement is for the lube oil pumps to take their suction from the lube oil tank and the hydraulic control valves to take their suction from a bearing header. The system may contain between 2000 and 10000 litres of oil.

The electrical engineer's interest in this system is that out of, say, six lube oil pumps on each generator set, five are driven by electric motors. The following is a description of the function of such pumps:

Main pump   The main lube oil pump is a shaft driven positive displacement unit, mounted into the inboard wall of the lower casing of the accessory gear. It is driven by a splined quill shaft from the lower drive gear and the pressure is 65 psig maximum. As the output pressure of this pump is engine speed dependent, with certain models of turbine insufficient lubrication pressure is available from the pump below a certain speed. In this case, even when it is operating satisfactorily, the pump may require to be supplemented by an electrically driven pump.

Auxiliaries pump   The auxiliaries pump, mounted on the oil tank cover, is a submerged centrifugal pump which provides lube pressure during startup and shutdown of the generator under normal conditions. The auxiliaries pump is driven by a low-voltage 30 hp two-pole AC flameproof motor.

Emergency pump   The emergency pump is also mounted on the tank cover, and is a submerged centrifugal type which also provides lube pressure under startup or shutdown conditions. This pump is driven by a 125 V 5 hp DC flameproof motor.

The three remaining pumps of our typical system are:

Main hydraulic supply pump, auxiliary hydraulic pump   Failure of these pumps and the resulting low hydraulic pressure would not necessarily cause an immediate generator set failure, although the unit would eventually trip owing to low lube pressure.

Hydraulic ratchet pump   Loss of hydraulic ratcheting pressure or equipment would cause generator shaft bowing and would also lead to excessive bearing stress. It may, however, be possible to repair the fault by, for example, replacing the motor within a few hours, i.e. before serious damage has been caused.

These three pumps would normally all be driven by low-voltage AC motors.

The basic criteria for establishing electrical supplies to generator lube oil pumps and other vital auxiliaries are as follows:

1. Although weight and space limitations usually prevent auxiliary switchboards being fed from separate generator transformers, as one would expect to find in a power station onshore, the switchboard or individual motor starters should be electrically as close to the generator as possible.
2. Each complete set of electrically driven auxiliaries should be supplied from one switchboard, so that only one supply is required to ensure that the particular set of auxiliaries is available.
3. Any standby auxiliaries should be fed from another switchboard, which obtains its supply from the generator via a different electrical route.

4.   If possible, the complete set of auxiliaries should be duplicated in order that a fault on one switchboard or its incoming supply will not lead to a shutdown of the generator.

5.   Loss or temporary disconnection of supplies to the emergency DC lube oil pumps should be adequately displayed on the generator control panel annunciator, so that the operator is aware that loss of AC power, and consequently the failure of the AC driven lube oil pumps, may lead to damage to the generator as it runs down.

3.9   Governors

Sophisticated electronic governors able to provide reliable service coupled with high performance have been available for some time. However, there are a few points worth considering when selecting a suitable governor. First, it is advisable to obtain as many of the governor's control system parameters as possible from the governor manufacturer. These will be required if the power system is to be computer simulated before the project design stage is complete, which is usually the case.

Any governor-engine-alternator system will have a finite response time. It also has a certain amount of stored energy which can be extracted to maintain the system by sacrificing some speed, and this energy may be used while the governor is responding to a load increase. Apart from large power system disturbances such as large motors starting or short-circuit faults, which need to be studied by dynamic simulation, any cyclic loads need to be considered carefully. The following anecdote illustrates the problem.

A radio transmitting station was built in an isolated location without access to any external electricity supply, depending entirely on two diesel generators for electrical power. The station was successfully commissioned, with the exception that the supply frequency became unstable and the generators shut down during each broadcast of the time pips. The problem was traced to the cyclic loading imposed by the series of pips being at a frequency at or close to the natural frequency of the governor mechanical linkages. In this case the problem was overcome by substantially increasing the mass of the engine flywheel. The only cyclic loads likely to be experienced offshore are those produced by large reciprocating pumps and compressors or by heaters which use thyristor integral cycle firing controllers. The phase angle controlled rectifiers and variable frequency inverters used on the drilling rigs are unlikely to produce the necessary supply frequency subharmonics.

3.10 Alternators and excitation systems

The control system parameters for the generator excitation system, like those for the governor, will be required if any computer simulation work is to be carried out.

The following aspects need particular attention in offshore systems:

1. Although, as with large onshore machines, two-pole 50 or 60 Hz designs are used because of the greater efficiency of energy transfer at the higher speeds, it should be remembered that this requires complex engineering analysis and the use of high-grade materials, particularly as the generator module could be located on a 200 m high steel structure, 30 m above sea level.

2. To the author's knowledge, the highest generator rated voltage offshore is 13.8 kV, and alternator manufacturers have no difficulty in producing machines at this rating. Voltages up to 22 kV could be used, however, provided that suitable switchgear is available.

3. The alternator subtransient reactance X_d'' is a useful regulator of maximum prospective fault current, and alternator manufacturers are usually prepared, within certain limits, to vary the design of the windings to enable the system designers to limit the prospective fault level to a value suitable for the switchgear available. For a machine of, say, 30 MW, the degree of variation for X_d'' would be approximately 15% to 21%.

4. Although, as discussed above, a highly reactive machine may be beneficial in order to limit prospective fault currents, this is accompanied by the penalty of poor motor starting performance due to the increased transient reactance (X_d'). The winding reactance must therefore be optimized for the best motor starting performance, allied with prospective fault capabilities within the capacity of the switchgear installed. This tradeoff is best accomplished using computer simulation.

5. The conventional configuration of brushless alternator with pilot and main exciter is commonly used offshore for machines of 500 kW rating and above. Machines with static (rectifier derived) excitation are acceptable provided fault currents can be maintained for at least the full generator fault time rating. A definite time overcurrent device should be used to shut down the generator within this time, so that the machine is not left running with a fault still on the system after the voltage has collapsed.

6. The automatic voltage regulator should incorporate a means of detecting a control loop disconnection such as that caused by open-circuit voltage transformer fuses, in order to avoid excessive voltages being developed on the machine stator if such a disconnection occurs.

3.11 Neutral earthing

Typical low-voltage solidly earthed and medium-voltage resistance earthed systems are shown in Figures 3.3 and 3.4. There is little difference between offshore and onshore practice with regard to generator neutral earthing. However, it is worth repeating that earth cables and earthing resistors should be adequately rated, both for current magnitudes and for circuit breaker tripping times. Because of the marine environment, earthing resistors tend to deteriorate quite quickly and require very regular maintenance (in the author's experience, at least once a year).

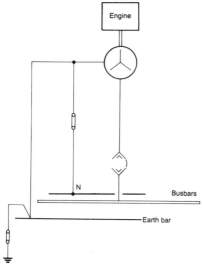

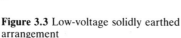

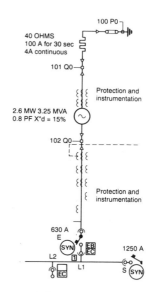

Figure 3.3 Low-voltage solidly earthed arrangement

Figure 3.4 Medium-voltage resistance earthed arrangement

3.12 Starting requirements

At some stage in the design of the platform power system it is well worth while carrying out a failure mode, effects and criticality analysis (FMECA) of the system to ensure that as many operational problems as possible can be foreseen and catered for in the design. FMECA methods are discussed in Chapter 12.

Consideration of operational problems is especially important in providing for the bringing of generators into service when all or most platform services are unavailable, i.e. black start facilities. Although written black start procedures should be available to the operators, these should reflect the permanent facilities installed. Black starting cannot be safely or adequately catered for by describing some temporary rig on the installation standing instructions. For example, the installation manager would not thank the system designers if in certain conditions it became necessary to fly a small generator set or compressor out to the platform in order to restart generators and hence continue the output of oil from the installation.

Maintenance must also be catered for in starting facilities such that whilst part of the system is being maintained, there is still at least one method of black start open to the operator. An example of this would be the need to allow for planned or unplanned outages in the emergency switchboard whilst still providing electrical supplies to main generator facilities, such as engine cranking motors and lube oil auxiliaries; this is a provision which has often been overlooked in the past.

3.13 Emergency generation

Most offshore production installations have three or four main levels of operation which are reflected in control systems such as the ESD system (see Chapter 5). If, however, there is a very large gas leak such that the installation is enveloped in a gas cloud, it would be necessary to isolate all forms of electrical power capable of igniting the gas, including in some instances the DC secure supply batteries. Assuming this dire situation has not occurred, the first level of operation is on battery power only and is considered in detail in Chapter 5.

The next level of operation is with the emergency generator only running. The provision of an emergency generator is a statutory safety requirement, and as such it should be designed to provide reliable power for statutory communications equipment, navigational aids, fire and gas monitoring, ballast systems and (although not statutory requirements) accommodation cooking, drinking water and sanitation facilities. As this generator must not be dependent on the platform production processes for fuel, it is invariably diesel driven. Storage of petrol or propane on the platform would be considered a hazard, which would rule out the use of an ignition engine for this purpose. These generators are usually designed to be automatically started on failure of other, larger generators on the installation. Again, there is a statutory requirement that the starting equipment for this generator is capable of at least six start attempts.

This generator should be located in a safe area, close to the accommodation, radio room and process control room. A day tank is required near the generator, big enough to run the machine for the time specified in the relevant statutory regulation. The time will vary depending on other installation conditions, such as whether it is regarded as manned or unmanned, but may be 24, 48 or even 96 hours.

The following points are often overlooked in specifications for emergency generator sets:

1.   Despite the small size of the prime mover, air intakes must still be provided with spark arresting devices and overspeed flap valves, and exhausts with spark arresters.
2.   Interlocking facilities must be provided to ensure that the generator circuit breaker cannot close on to an existing fault when the generator is automatically started.
3.   Means should be provided to maintain the generator output current in the event of a fault, for long enough to operate protection devices, where this is possible with the limited magnitude of fault current available from such a machine. Leaving the machine running with collapsed excitation is dangerous, as the fault may disappear, to be followed by a sudden and possibly unexpected reappearance of full voltage on the system. The photograph in Figure 3.5 shows a typical current design of an emergency diesel generator set.

Figure 3.5 Typical current design of emergency diesel generator set. (Courtesy of SPP Offshore, a division of SPP Ltd.)

3.14   Key services generation

For intermediate operating conditions, for example when certain process or service machinery is needed but the level of production does not require the running of main generators, key services or submain generators may be run. These would allow all the utilities, such as ventilation and cooling systems, plus the statutory services, to be operated without the need to run the main generators at inefficient power levels. Key services generators can also be used to provide peak load power, particularly during a planned outage of a main generator. It is best, however, to avoid running machines of different sizes in parallel, since the shorter time constants usually associated with the smaller machine tend to cause it to react more quickly to step increases in load, and this leads to system stability problems.

Chapter 4

Process drives and starting requirements

Machinery drives for offshore installations range in size from fractional horsepower cabin ventilation fans to the 15 MW or more machines required for gas injection compressors. Some concern has been expressed over the practicality of installing very large electric motors offshore owing to the stresses imposed on the structure, especially during starting. A figure of around 25 MW depending on the application (e.g. less for reciprocating pump drivers) should be considered as a ceiling figure for offshore electric motor ratings.

4.1 Voltage levels

As with onshore industrial plant, motor operating voltage levels correspond to ranges of motor power ratings, in order to keep current and voltage drop magnitudes within practical limits. For weight saving purposes, the step points at which higher operating voltages are selected tend to be lower offshore, as follows:

Voltage level	Rating
415 to 460 V	up to 150 kW
3300 to 6600 V	up to 1000 kW
11 to 13.8 kV	over 1000 kW

Higher voltages such as 22 kV could be utilized provided that the economics are favourable and that proven ranges of machine at these higher voltages are available.

4.2 Starting

Where an individual motor load represents a substantial part of the power system capacity, it is desirable to obtain designs of squirrel-cage motors with the lowest starting current characteristics compatible with the driven equipment. Both reduced voltage starting of cage motors and the use of slip ring motors will add considerable weight to the drive package, and therefore direct-on-line (DOL) low-starting-current machines are preferred unless technical considerations for selection are overridden by those of cost and delivery. Motors with starting currents in the region of 3.5 to 4.5 instead of the normal 6 to 10 times full load are usually available from the larger manufacturers.

4.3 Speed

The shaft speed of a pump, compressor or fan is critical to its performance. It is therefore necessary to obtain the required speed or speed range either by installing a gearbox between the motor and the driven equipment, or by selecting a suitable system frequency and pole configuration for the motor.

In the 1930s the US Navy changed from DC to AC systems and selected 440 V, 60 Hz as the operating parameters. Since then the whole of the NATO fleet and the majority of commercial ship designs have standardized on these parameters. This has greatly assisted in improving the availability of 60 Hz options on the standard equipment ranges of most European manufacturers. Although with small low-voltage motors the increased efficiency due to higher pump speeds at 60 Hz is marginal, with motors having ratings of the order of several megawatts the weight and power savings can be substantial.

With smaller installations, however, operating at 60 Hz may be a disadvantage where it is decided to select a reciprocating engine rather than a gas turbine main generator prime mover. The problem is that the optimum engine speed of around 1500 rpm is better suited to generating at 50 Hz. Reciprocating engines running at 1200 rpm tend to have too low a power-to-weight ratio, and operating at 1800 rpm leads to short cylinder life or even piston speeds which would be beyond the design limitations of the engine.

The higher synchronous speeds obtainable at 60 Hz also lead to higher inherent noise levels, although this can be deadened with better module insulation.

From a machinery standpoint, a major disadvantage in adopting 60 Hz for an offshore installation is related to testing the equipment prior to installation on the platform, since full load tests cannot be carried out using the British and European national supply networks. Until recently, tests were mainly carried out at 50 Hz and the results extrapolated to give projected machine characteristics at the design conditions. However, test facilities are now available in the UK for motors of up to 6 MW at 60 Hz. For larger machines, where capital investment is high and full load tests are considered essential, it is usually possible to arrange full load tests in conjunction with testing of the main generators to be installed on the offshore installation. Although this procedure is usually expensive, the costs should be more than offset by the benefits of adopting the higher frequency.

Once the system frequency has been established, it becomes increasingly expensive to change and hence may no longer be considered a variable after that point in the system design. It is therefore important to consider the number, rating and purpose of the larger drives on the installation at an early stage in the power system design, before the frequency is selected.

However, in many cases the shaft speed available from a motor, even from a 60 Hz two-pole machine (i.e. 3600 rpm), is lower than the required shaft speed, and it will still be necessary to install a gearbox in the drive string. Once the requirement for a gearbox has been established, changes in the drive ratio have only minor effects and the motor speed may be chosen to give the optimum motor design in terms of dimensions, weight, reliability, noise emission and so on.

4.4 Pole configuration

At a given frequency, motor synchronous speed is determined by the number of poles incorporated in the motor stator, and this governs the maximum operating speed of the machine. The fastest possible speed of both induction and synchronous motors is with a two-pole configuration, which gives synchronous speeds of 3000 rpm at 50 Hz and 3600 rpm at 60 Hz respectively.

As size and weight penalties are usually incurred by increasing the number of poles, only two- and four-pole machines are normally used offshore. An occasional exception is large reciprocating compressor drivers, where eight-pole (or more) synchronous machines may be used. These machines, apart from giving the required lower speed, also reduce the current fluctuations caused by the cyclic torque variations associated with reciprocating machinery. The advantages and disadvantages of two-pole against four-pole machines are as follows.

4.4.1 Advantages of two-pole machines

1. Energy conversion within the higher-speed two-pole machine is usually more efficient, giving some reduction in size and weight for a given output. However, because of the higher rotor speeds, especially at 60 Hz, mechanical forces on the rotor cage, known as hoop stresses, become significant and limit the maximum dimensions of rotor that may be manufactured using conventional materials. Metal-lurgically more exotic materials may be used to extend this limit with, of course, the accompanying large increases in the cost of the machine.
2. Manufacturers differ in the application of a practical maximum rating limit for a two-pole motor, but as a general rule this is between 3 and 5 MW. Therefore motors above 5 MW should not be considered as a feasible alternative unless the cost of using exotic metals in the rotor is outweighed by the savings accrued by, for example, the elimination of a gearbox.
3. Below this limit, and within manufacturers' normal product ranges, the use of a two-pole machine should, in comparison with an equivalent four-pole machine, provide dimension and weight savings roughly proportional to power rating; however, for small machines of only a few kilowatts there would be little benefit.

4.4.2 Disadvantages of two-pole machines

1. Less starting torque is available from two-pole machines, requiring the driven equipment to have a lower moment of inertia. It may also prove more difficult to accelerate the machine up to operating speed, where driven machinery cannot be run up to speed unloaded. The speed/torque characteristics of pumps in particular should be carefully studied to avoid any problems. Starting currents are also likely to be higher, and with large machines this may lead to unacceptable voltage dips.
2. Irregularities in the core stampings, which are inevitable unless very high levels of quality control are applied, generate more magnetic noise in two-pole machines. A characteristic low-frequency (twice slip frequency) growl can be heard from these motors.
3. Rotor imbalance is more likely and can cause more vibration on two-pole machines.
4. The higher rotor speeds associated with two-pole machines will shorten bearing life or require a more expensive higher-performance bearing.
5. Poorer heat dissipation within the rotor necessitates increased cooling air flow rates in two-pole machines. The effect of higher fan speeds and increased air flow rates is to increase noise emission from the machine.

From the above, it can be seen that the two-pole machine generally scores on weight and dimensions but suffers from the design limitations associated with higher speeds. Certain applications such as direct drive ventilation fans and axial compressors can often take considerable advantage of the higher-speed machine. Nevertheless, if no significant benefit is obtained from the higher speed then the four-pole machine should be used.

4.5   Cooling and ingress protection

As discussed in earlier chapters, only fully enclosed types of motor are normally considered suitable for offshore installations. Three typical enclosure types are to be found offshore, as now described.

4.5.1 Totally enclosed fan ventilated (TEFV)

In this motor type, the stator windings are enclosed within a finned motor casing. Cooling is achieved by the mounting of a fan on the non-drive end of the rotor shaft, external to the stator end-plate, in order to blow air over the external cooling fins. The fan fits inside a cowling which deflects the air over the fins. An ingress protection rating of IP55 is typically achieved with this design, which provides a high level of physical protection. The majority of low-voltage machines, especially in hazardous areas, are of this type.

The chief disadvantage with this type of motor is the inherently high noise level produced by the external fan. Methods are, however, available to reduce this noise, such as the use of acoustically treated fan cowls, or the use of a machine oversized for the application but fitted with an undersized fan.

This design is unsuitable for the larger medium-voltage machines, but can be used advantageously with the smaller 3.3 or 4.16 kV machines.

4.5.2 Closed air circuit, air cooled (CACA)

As motor sizes increase, it becomes less and less likely that heat generated in the rotor and stator windings can be dissipated to the machine casing at a fast enough rate to prevent design insulation temperature rises being exceeded. It then becomes necessary to pass a cooling medium (usually air) through the inside of the motor to remove surplus heat. With the majority of CACA types found offshore, air is usually forced through the windings by a rotor mounted fan within the motor casing. Since the machine is still required to be enclosed, this cooling air is recirculated in a closed circuit through the machine, and heat is extracted by an air-to-air heat exchanger mounted on top of the motor. A second rotor driven fan is often required to force the external air over the heat exchanger.

The CACA design suffers from the same noise and weight problems as the TEFC type, and the air-to-air heat exchanger only adds to weight and bulk. Nevertheless it does provide an adequate and simple method of cooling the larger motors and – important from the reliability point of view – requires no external services in order to continue operating.

4.5.3 Closed air circuit, water cooled (CACW)

If the air-to-air heat exchanger on the CACA machine is replaced by an air-to-water unit, we then have a CACW machine. The machine is dependent on an adequate supply of cooling water for continued operation. The benefits of this arrangement are that the bulk of the heat exchanger is very much reduced, and there is no requirement for the secondary cooling fan and consequently there is a substantial reduction in noise.

The disadvantages are first that the machine depends on the cooling water supply, and hence there is a reduction in reliability; and secondly that the presence of water under pressure around the machine is a hazard.

To summarize the above, the following is recommended for enclosure selection:

TEFC Smaller low-voltage machines.
CACW Where the TEFC design is not practicable, i.e. with larger 3.3/4.16 kV and all higher-voltage machines.
CACA Larger machines where the cooling water supply is uneconomic or the machine must operate during a cooling system outage.

4.6 Special applications

This section provides some advice on the selection of motors for particular applications.

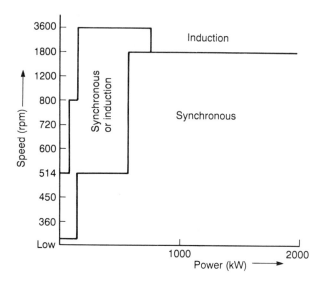

Figure 4.1 Ratings of induction and synchronous motors for compressor drivers

Hazardous area topics are discussed in Chapter 8 and will not be presented in detail here. However, readers who are likely to be specifying motors for hazardous areas would be advised to read Chapter 8 next (see also Figure 4.1).

4.6.1 Reciprocating pumps and compressors

If an induction motor is used to drive a large reciprocating pump or compressor, the heavy cyclic torque fluctuations demanded from the motor will in turn demand heavy current fluctuations from the supply. When the motor load is a significant part of the installation generating capacity, instability of voltage and power may result.

An alternative is to use a synchronous motor with a squirrel-cage damping winding embedded in the rotor. If a steady torque is being developed by the machine, the load angle would remain in an equilibrium position. Since the rotor of a synchronous motor running in synchronism with the supply experiences a torque proportional to its angular displacement from the equilibrium position and also possesses rotational inertia, it constitutes an oscillatory system similar to the balance wheel of a clock. If J (kg m^2) is the moment of inertia of the rotor, then it can be shown that the natural frequency of the rotor will be:

$$f = \tfrac{1}{2\pi} \sqrt{T_s/J} \times (\text{No. of pole pairs})$$

where $T_s = 3VI/\cos\theta$ N m, V is the system voltage, I is the current produced by the field induced voltage, and θ is the load angle.

Synchronous motors driving reciprocating machinery receive torque impulses of a definite frequency, and for satisfactory operation the natural frequency of the rotor must be at least 20% higher or lower than the frequency of the torque impulses (Figure 4.2).

The embedded squirrel-cage damping windings, used for starting, will produce some corresponding current fluctuation with torque, but this is not excessive as can be the case with an equivalent induction motor. Such windings produce damping torques proportional to the angular velocity of any rotor oscillation, and hence reduce the synchronous motor's tendency to hunt due to alternating currents induced in the other windings and current paths of the rotor, giving rise to destabilizing torques.

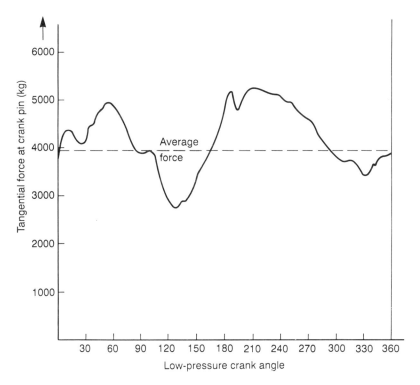

Figure 4.2 Typical torque requirements over one cycle of reciprocating compressor operation, full load

4.6.2 Gas compressors

It is worth remembering that large gas compressors, whether using induction or synchronous motors, are very dependent on their auxiliaries for reliable and safe operation. The following is a list of basic requirements:

(a) hazardous area ventilation fans;
(b) main and standby lube oil pumps;
(c) main and standby seal oil pumps;
(d) cooling water pumps;
(e) motor ventilation fans.

This list is not exhaustive, as it does not include the various installation utilities and safety systems which have to provide continuous permissive signals to allow starting or continued running of the compressor package.

4.6.3 DC drilling motors

The conventional arrangement in a drilling rig is to utilize 750 V DC machines run from phase angle controlled thyristor units. Typically the machines must be capable of accepting a voltage variation of 0–750 V DC and continuous load currents of 1600 A. For mud pump duty, two motors will run in parallel on one SCR bridge. For the draw works and rotary table, one SCR bridge will be assigned to each motor with appropriate current limiting devices in operation. The motors must also be capable of producing around 600 kW continuously and 750 kW intermittently at 1100 rpm. These machines are shunt wound machines with class H insulation, derived from railway locomotive designs. However, as they operate in hazardous areas, the construction is closed air circuit, water cooled (CACW) in order to restrict surface temperatures; the enclosures are pressurized to prevent ingress of explosive gas mixtures.

This type of motor is often provided without a terminal box; the winding tails pass through a sealed and insulated gland to a separate flameproof terminal box. It is recommended, however, that motor mounted terminal boxes should be used, since the exposed winding tails are difficult to protect mechanically.

4.6.4 Power swivels

On some drilling rigs the drill string is powered by the swivel instead of the rotary table. The system consists of the swivel powered by a hydraulic motor fed by hoses from a hydraulic power pack which is located in a pressurized room. The power pack consists of a swashplate pump driven by a medium-voltage motor. The motor is a conventional squirrel-cage type as described above.

4.6.5 Sea water lift pumps

In floating installations or those with hollow concrete legs, conventional pumps may be used to obtain source water for cooling since the pump may be located at or near sea level. However, on steel jacket construction platforms it is necessary to draw sea water for cooling and fire fighting up to the topsides using sea water lift pumps.

There are three basic types of electric drive for sea water lift pumps:

(a)   submersible electric;
(b)   submersible hydraulic;
(c)   electric shaft driven.

Types (b) and (c) use standard forms of electric motor as the power source; power is transmitted mechanically or hydraulically to the location of the pump (see Figure 4.3). Therefore only type (a) will be discussed as a special application.

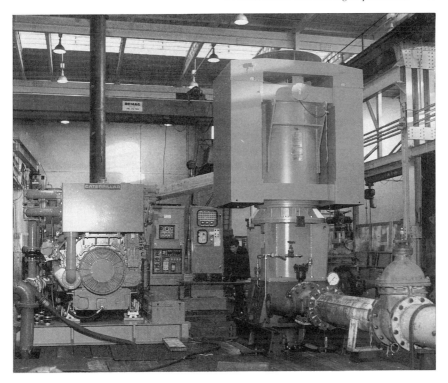

Figure 4.3 Typical electric shaft driven fire water pump. (Courtesy of SPP Offshore, a division of SPP Ltd.)

Submersible pumps for this duty are of small diameter and usually very long, consisting essentially of a series of small induction motors all mounted on the same shaft. The motor-pump string, having been connected to a special flexible cable, is lowered down the suction or 'stilling' tube. Alternatively, the stilling tube may be sectionalized and the motor fixed in the lowest section of the tube. The tube is then lowered down a platform riser and jointed section by section until the motor suction is 10 metres or more below the sea surface at the height of the lowest expected tide. For cooling purposes the motor must be below the pump, and therefore difficulty is often experienced in avoiding damage to cables which have to pass between the pump and the stilling tube in order to reach the motor terminals. Bites need to be taken out of the pump retaining flanges to allow the cables past, and the cable overall diameter must allow a loose fit through these to avoid damage. Photographs of this type of pump set are shown in Figure 4.4(a) and (b).

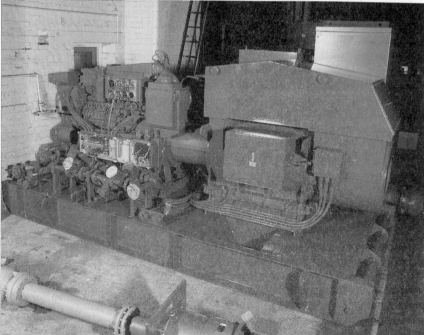

Figure 4.4 (a) Diesel-electric fire pump. The pump (foreground) is installed in a vertical submerged stilling tube; (b) rear view. (Courtesy of SPP Offshore, a division of SPP Ltd.)

4.6.6 Diesel-electric fire pumps

Statutorily, every offshore installation has to be provided with at least two (depending on the capacity) serviceable fire pumps, each of which must be powered independently of the other. A third pump must be provided to cater for unavailability during servicing. Pumps must also be physically segregated and located geographically well away from each other so as to minimize the risk of both pump systems being damaged by the same fire or explosion. Further details on capacities of pumps etc. are obtainable from the Department of Energy's *Offshore Installations: Guidance on Firefighting Equipment*.

A typical fire pump arrangement consists of one motor driven pump powered from the installation power system, and two pumps driven by dedicated direct diesel or diesel generator sets. If a diesel generator set is to be dedicated to the supply of a particular submersible pump, weight and cost may be reduced by dispensing with any switching device and, in the manner of a diesel-electric locomotive, cabling the generator directly to the pump. The usual arrangement is to use a fire survival cable from the generator to a terminal box at the top of the stilling tube, where it is connected to the flexible stilling tube cable. The advantage of this arrangement is that the motor starts with the generator, effectively providing a reduced voltage start characteristic. A generator of lower rating than that required if the motor were started from a switching device can then be used. Another benefit of this arrangement is that a generator voltage can then be specified which minimizes the stilling tube cable diameter, i.e. copper and insulation cross-sections can be traded off. The fire pump diesel generator module is designed to be as independent as possible from other platform systems, and if using cooling water bled from the fire pump system itself, only requires to obtain combustion air from its surroundings. The photograph in Figure 4.5 shows a typical diesel generator module of current design.

4.6.6.1 Fire pump diesel engine starting requirements
The starting requirements for fire pump diesel engines are laid down in the (US) National Fire Protection Association (NFPA) specification 20. This requires that the engine be provided with two batteries each capable of 12 start attempts, specified both for minimum cranking duration and for interval between each cranking. The engine must also have an independent means of starting, such as a hydraulic or compressed air device complete with some form of accumulator.

4.6.7 Downhole pumps

Downhole pumps are used in the latter stages of well life to provide artificial lift when wellhead pressure or crude oil flow rates need to be improved. Downhole pumps are extremely rugged devices, having to work in a 'hole' less than 0.25 metres in diameter, hundreds of metres below the sea bed, and at pressures and temperatures near the design limits. The

Figure 4.5 Typical fire pump/emergency generator module designed for fire survivability and independent operation. (Courtesy of SPP Offshore, a division of SPP Ltd.)

pump-motor string, although of very small diameter, is often in excess of 10 metres long.

The pumps are expensive proprietary devices; their manufacturers closely guard their design secrets. The installation of a downhole pump may cost in excess of £2 million; as usually no warranty is provided, that sum may need to be spent again a few days after installation if nothing happens when the start button is pressed!

However, much can be done at the surface to improve the reliability of the pump. Notably, the use of variable frequency converters to provide a very soft start for the motor has proved to be very successful in the last few years. Downhole pump motor nominal surface voltages are usually around 2 kV to allow for voltage drop in the cable and to trade off copper and insulation cross-sections to minimize motor dimensions. As the pump may be capable of lifting in excess of 20 000 barrels per day of oil and water from the well, the frequency converter and transformer units tend to be large.

4.6.8 Main oil line (MOL) and water injection pumps

Apart from the hazardous area requirements, the motor drivers for these pumps will be very similar to those used onshore. Owing to the pressures involved (typically 200–400 barg) care needs to be taken in the design of controls to ensure that no undue stresses are put on the pump or pipework, particularly during startup. It should not be possible to start the motor if the pump-motor set is running backwards owing to the wrong valves being inadvertently opened. As well as causing shock to the pipework, such maloperation may draw starting currents of excessive magnitude from the system, possibly causing damage to the motor windings and/or power system instability.

Chapter 5

Control and monitoring systems

5.1 Generator controls

With any electrical system, the importance of ergonomically designed controls cannot be overstressed. The following controls and instrumentation are considered to be the basis for a generator control panel. The logic of a typical generator control panel is shown in Figure 5.1.

5.1.1 Controls

5.1.1.1 Start and stop buttons
The start control normally has the function of initiating the engine automatic start sequence and, depending on the control philosophy adopted, may automatically synchronize the generator with any generators already on load. An auto/manual selector switch may also be provided. This is useful during commissioning, to allow an individual check of each step in the sequence to be made.

Two stop controls are often provided. One is the normal stop button, which initiates a timed rundown of the generator load and allows the engine to cool down before it is stopped. In the case of large gas turbines, a ratcheting sequence will also be required. This is a facility whereby the engine is rotated at intervals to prevent hot spots developing which may cause the misalignment of the main shaft.

5.1.1.2 AVR and governor raise/lower switches
These controls are used to set the voltage and frequency of the generator and, when it is in parallel with another generator, to allow the sharing of reactive and real power to be adjusted respectively. It is an advantage to group these controls and associated electrical metering such as voltmeters, kW meters and kVAr meters so that manual adjustment of real and reactive power can be carried out by one operator. If the controls are spread across four or five panels, each associated with an individual machine, adjustment can be difficult. If there are only two machines, then controls can be mirrored so that raise/lower controls are located close together in the area on the panels near where they butt together.

An auto/manual voltage control selector switch is also required, so that commissioning and routine checks may be carried out on the excitation system. In some systems, it may not be possible to switch easily from auto to manual voltage control or vice versa. As AVRs are now solid state devices and take up very little room in the control panel, a dual AVR system is recommended for all but the smallest machines. With a dual AVR system, a standby AVR follows the main AVR and automatically takes over if the main unit fails.

5.1.1.3 Synchronizing equipment

It is well worth providing a good selection of synchronizing controls and indicators on the panel, as this not only provides for safer operation but also gives the operator more confidence during paralleling operations. An auto synchronizing unit (auto synch) and a check synchronizing relay (check synch) should be provided; the auto synch is normally used with automatic sequencing and the check synch when any paralleling is being carried out manually. The function of both devices is to provide a 'close permissive' signal to the generator circuit breaker when voltage and frequency conditions are suitable for paralleling. If the signal is not present, circuit breaker closing is prevented by the control logic.

5.1.2 Instrumentation

5.1.2.1 Metering

The control panel should provide at-a-glance information to allow the operator to carry out his work. Voltage and frequency displays should be digital and preferably to no more than two decimal places for fast reading. If the generator nominal voltage is more than 1 kV, the display should also be in kV, again for fast reading. Although this means that the operator only sees changes in steps of 10 V, this is better than can be seen from an equivalent analogue panel meter and is compatible with the accuracy of the transducer. Real and reactive power meters should preferably be analogue, as they are used mainly for balancing loads between machines and the pointer semaphore effect is more quickly appreciated by the operator. A power factor meter should be provided of either digital or analogue type; in either case, the words 'leading' and 'lagging' should be displayed rather than '+' and '−'.

5.1.2.2 Synchronizing indicators

The indication provided by a particular instrument should preferably be duplicated by another for safety and operator confidence. For instance, a synchroscope should be supplemented by synchronizing lamps (preferably three) for manual synchronizing. Lamps should also be provided to indicate that check and auto synch permissive signals are present during circuit breaker closing. Where generator controls cannot be grouped conveniently, it will be necessary to provide busbar voltage and frequency meters as an aid to the operator during paralleling.

If there are more than two generators in the system, a mimic panel is recommended. The mimic should show the electrical system in diagram-

54

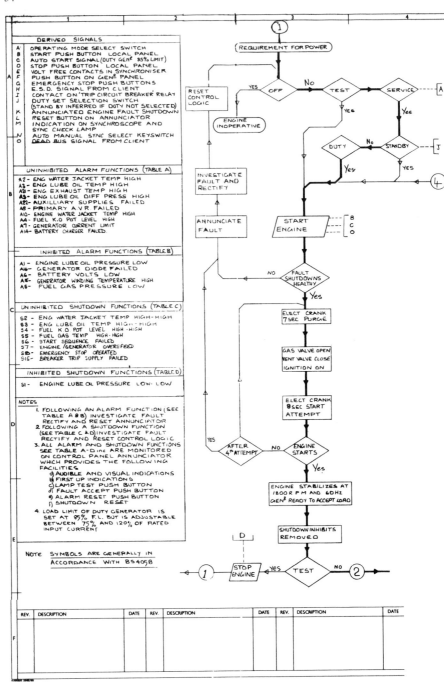

Figure 5.1 Logic diagram for a typical generator control panel

55

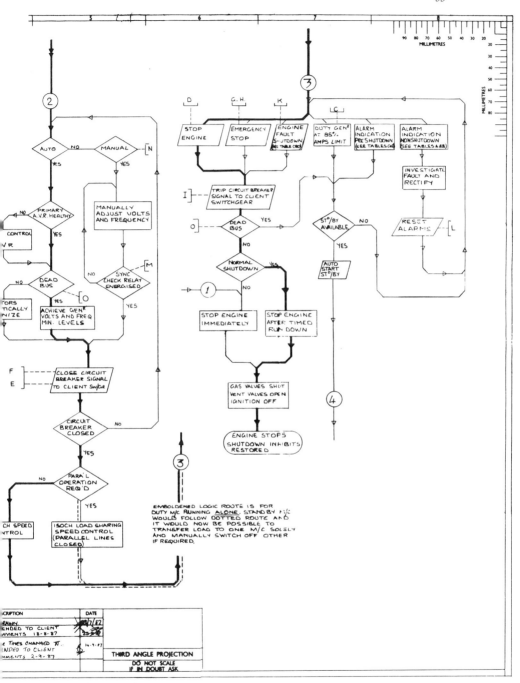

EMBOLDENED LOGIC ROUTE IS FOR
DUTY M/C RUNNING ALONE. STANDBY M/C
WOULD FOLLOW DOTTED ROUTE AND
IT WOULD NOW BE POSSIBLE TO
TRANSFER LOAD TO ONE M/C SOLELY
AND MANUALLY SWITCH OFF OTHER
IF REQUIRED.

THIRD ANGLE PROJECTION

DO NOT SCALE
IF IN DOUBT ASK

matic form. The various controls and indicators for each machine should be located on the mimic at positions where they are clearly associated with the machine whose parameters they are displaying or controlling.

5.1.2.3 Alarm annunciator

As with any machinery, it is necessary to monitor each generator for faults and failures and bring them quickly to the attention of the operator. This is usually accomplished by using a matrix of windows, each engraved with a particular fault or failure. Each of these windows will be lit when a particular fault is sensed by a transducer located on the engine or generator. The transducers may be level switches, for example monitoring lube oil level and operating on low oil level to light up the annunciator engraved 'low oil level'. The logic of the annunciator should also indicate to the operator which fault occurred first. This first-up facility is usually provided by making the window of the first fault flash on and off, or by making it remain steady and any later faults to flash. The occurrence of any alarm or shutdown will be accompanied by an audible alarm, which will be silenced when an 'accept' button is pressed on the panel. There should also be a lamp test button which while pressed causes all the lamps on the matrix to light up. Once the fault has been cleared, the annunciator can be cleared of all fault and failure indications by pressing the reset button.

Annunciator windows should be segregated into groups dealing with different types and severities of fault. Faults which cause an immediate shutdown must be indicated by windows of a different colour and be well segregated, usually by being placed on the lower lines of the matrix. Recommended colours are amber for faults which do not cause an immediate shutdown and red for those that do.

5.2   Load sharing systems

The operator's workload may be lightened by introducing facilities which automatically control the sharing of two or more generators running in parallel.

The principles of a typical load sharing system are as follows. An astatic loop control system is set up between the AVRs of each generator. This operates by comparing a DC reference voltage with DC voltages derived from respective AVR control signals. The power sharing is accomplished by monitoring the output power of each machine and feeding this to an associated power comparator unit. Each machine power comparator also receives a signal from a frequency controller unit. This compares the actual supply frequency with a frequency reference, so that the resulting comparator output signal is a function of the generator output modified by any frequency error. The outputs of each machine power comparator are linked to the next, forming a comparator loop which provides each associated governor with an appropriate power mismatch signal. The mismatch signal is used to drive governor raise/lower relays. The resulting system, once set up correctly, will provide good load sharing of both real and reactive power over a wide range of system loads.

5.3 Load shedding systems

As with most electrical supply systems, system loads tend to grow as more and more equipment is installed on the platform. Rather than install another generator set, which may be an unnecessarily drastic solution to the problem, it may be possible to automatically disconnect non-critical equipment from the system until the peak demand period is over. This may be achieved very simply by using a trip setting kW meter which trips certain load contactors when a preset load is reached, or there may be a very sophisticated microprocessor controlled system with a multiplicity of inputs and outputs. In general, as the power output characteristics of most engines are sensitive to variations in ambient air intake temperature, some account of this should be taken when designing load shedding systems (see section on turbine temperature limits in Chapter 3).

5.4 Power management systems

With the advent of microprocessors, much more of the power generation system can be automated. To avoid embarrassing failures, microprocessors must be duplicated or triplicated so that errors may be detected and the system 'frozen' in a safe state so that operators may take over the function manually. The software used must also be extremely reliable, with built-in error trapping routines and means for ensuring a fast return to normal operation after a system crash. Memory devices should be read-only where possible; where read-write memory is necessary, this must be well protected from electromagnetic or other harmful effects. Any volatile memory must be permanently battery backed. Most microprocessor systems incorporate watchdog and visual indication on the occurrence of a fault.

Such a system, when properly programmed, can provide a whole range of useful facilities to make the operator's life easier, as follows:

1. Automatic starting and paralleling of generators as demanded by rising load and engine intake temperature according to a prearranged programme.
2. Automatic shutdown of surplus generators on falling load requirement or ambient temperature according to a prearranged programme.
3. Staged removal of non-critical loads from the system in order of priority, the number of stages being dependent on the severity of the overload condition.
4. Automatic startup and paralleling of another generator on occurrence of serious fault or failure of a running generator. This facility can easily be provided in a system with only two machines, but requires a programmable device where more than two machines are installed.

5.5 Fire and gas considerations

The prevention of fires and explosions in petrochemical plants is always a vital necessity. Offshore, where the sea can be as hostile an environment on a stormy winter's night as the fire, immediate evacuation of the installation should be only one of a number of alternatives available to the offshore installation manager, depending on the particular scenario.

The basis of any offshore fire and gas system is first to continuously monitor those areas of the installation where experience has shown the most probable fire or explosion risks occur; and secondly to automatically indicate to installation staff the whereabouts of the occurrence, initiate audible and visual alarms, and discharge extinguishing agents where the location has a permanent fire fighting system installed.

The fire and gas control panel is located in a non-hazardous area. It is usually manned continuously; the exception to this rule is if the platform is completely unmanned or if control is transferred to a shore terminal or adjacent platform during the night. Indication and control are then provided by telemetry facilities. The fire and gas panel will normally contain two sections: monitoring systems and extinguishing systems.

5.5.1 Monitoring systems

All fire and gas detectors, break-glass fire buttons and fire main flow switches will be monitored from this section. Different types of detector may be used depending on the location and the source of the hazard being monitored. The principle is to ensure that the particular hazard is detected as quickly as possible.

5.5.1.1 Fire detectors
Fire detectors used offshore are usually of the following four types:

1. Ionization smoke detectors, used in switchrooms and in accommodation areas.
2. Ultraviolet and/or infrared flame detectors, used in process areas to detect the flames of burning gas leaks. These should be shielded or filtered if used in sunlight or near vent flares, and may give spurious alarms if not inhibited during local welding operations.
3. Heat detectors, used in engine compartments and hoods to detect fires in the associated high ventilation-air flow rates where other detectors may be unsuitable.
4. Quartz bulb detectors, used in conjunction with sprinkler or deluge systems to release water on to pipework or in accommodation cabins when heat causes the quartz bulb to burst due to the expansion of the liquid inside.

5.5.1.2 Gas detectors
Most gas detectors used offshore employ a conducting element coated with a catalyst material, which causes the element to conduct a greater DC current in the presence of a hydrocarbon gas. Although these devices include voltage regulation circuits, it is advisable to provide a DC supply well within the tolerances specified by the gas detector manufacturer, and to check battery voltages regularly, with chargers switched off, to ensure the correct voltage is available. Low voltages or sudden voltage changes may cause spurious alarms. This type of detector can become poisoned by

the presence of chemicals such as paints and aerosol sprays which act in a similar manner to hydrocarbon gas, leaving the device continuously in the alarm state. A buildup of salt and grime may prevent gas reaching the gas sensitive part of the device so that it fails to an unsafe state. Such failed detectors must be found by regular testing and replaced. Infrared gas detectors have recently become available which can detect a range of gases and are impervious to poisoning. Open path types will monitor over some distance and are useful in external situations. It is necessary, however, to clean the lenses on the devices regularly.

Gas detectors are placed at strategic positions in process areas at points such as wellheads, separators and gas compressors, near flanged joints and in other places where leaks are comparatively more likely. Ventilation air intakes must all be monitored, especially those for non-hazardous areas such as accommodation where most electrical equipment will not be designed to prevent ignition, and where smoking is normally permitted. Gas detectors are not very fast acting (catalytic type more than 10 seconds, infra-red type 5 to 8 seconds) and also, for reasons given above, require frequent testing and maintenance to ensure a high probability of detection should a gas leak occur. For this reason, gas detectors are normally provided in pairs or sets of three or more, with 2-out-of-N voting logic. Gas detectors are also required in the intakes of engines to detect the presence of explosive mixtures in the intake ducting and, in the case of a turbine, the turbine hood. Detection of gas in such locations should lead to immediate shutdown of the engine in order to prevent damage due to explosion or overspeed. However, ingestion of gas from process leaks by governed diesel engines is unlikely to cause an overspeed, owing to the need for compression ignition (see Chapter 3).

5.5.1.3 Monitoring circuit logic
Combinations of different types of fire detector are used to obtain optimum cover for a particular situation. For example, a gas turbine hood will usually have a combination of heat and ultraviolet detectors. To improve reliability and to avoid spurious operation, several detectors are often operated in a voting arrangement so that, for example, one detector being activated will give an alarm whilst two or more will discharge extinguishant. As mentioned above, for gas detectors the 2-out-of-N logic is preferred.

Detector circuits are supervised for continuity and short-circuits by supervision circuits, and all monitoring logic operates on the fail-to-alarm condition principle, i.e. the occurrence of any failure which the monitoring systems are designed to detect will initiate an alarm. The cabling for detector circuits should be routed back to the panel so as to prevent one fire from disconnecting all the detectors in one particular area.

5.5.2 Extinguishing systems

The detection of fires in wellhead, drilling and production areas will cause the fire and gas panel to automatically respond by opening deluge valves in the area affected. For generator sets, it is a statutory requirement that installations over 750 kVA are provided with fixed fire fighting systems. The extinguishant normally used is a gas called halon 1301. However, as halon 1301 is a CFC, the use of other extinguishant gases is under discussion. If two generators are rated at less than 750 kVA but their total capacity is greater than this figure, the requirement will still apply unless the machines are separated by continuous fire walls without access doors or other openings. It is necessary to make the generator room as gas tight as possible in order to maintain sufficient halon concentration for the time it requires to extinguish the fire (at least 30 seconds).

In the case of a reciprocating engine, the engine should be stopped before the release of halon, as its air intake is normally from the engine room itself and the engine would scavenge the halon and reduce the concentration. In the case of a diesel engine, the engine would not be stopped by ingesting halon but would continue to burn, producing green exhaust smoke! To make the room gas tight and also to delay the spread of a fire into surrounding areas, the room will often be fitted with fire dampers which are normally held open by latches and automatically released prior to halon discharge, on detection of fire. If a manually activated system is used, then the stored energy required to release the fire dampers may be obtained from the halon bottles or from a compressed nitrogen bottle.

5.6   Process shutdown considerations

In order to prevent disastrous fires and explosions should part of the process plant malfunction, the emergency shutdown system interconnects the process controls in such a way that the whole process is automatically shut down to a safe level of operation, either through manual initiation by the operator or by safety instrumentation. The logic of the shutdown system consists of hierarchical levels such as the following.

5.6.1 Level 1

A level 1 shutdown would occur when an upset condition on a single unit requires that unit to be shut down. At this level, no other unit is affected in such a way as to necessitate its shutdown and the whole process should recover to stable operation.

For example, if one of the operating main oil line pumps trips, the system can continue to run; the wellhead flow control system automatically reduces throughput.

5.6.2 Level 2

A level 2 shutdown would occur if an upset in one system directly causes the shutdown of one or more other systems. This level of shutdown would cause the loss of production from one production train. An example of this would be the following automatically initiated series of events:

1. Separators block up.
2. All wellhead wing valves are shut.
3. All pipeline pumps are tripped.
4. Gas compression is shut down and depressurized, including the fuel gas system, if no other train is operating.
5. Sea water injection pumps are tripped.
6. Main generation gas turbines are changed from gas to diesel fuel operation.
7. Separators are depressurized and, after a time delay, drained.

5.6.3 Level 3

Level 3 shutdowns are initiated by external non-process conditions, or by an emergency which prevents safe operation of the production systems. Thus the sphere of control of a level 3 shutdown would be limited to systems directly involved with the production of hydrocarbons and to systems dependent on the continuation of production, such as water and gas injection.

An example of such a shutdown is as follows:

1. Wing valves, master valves and subsea valves are closed on production, gas and water injection wells.
2. Gas compression systems are shut down and depressurized, including fuel gas systems.
3. The separation train is stopped, depressurized and, after a time delay, drained.
4. The main generators are changed over from gas to diesel fuel.

Should the main and key services generators fail due to fuel supply disturbance or any other reason during this level of shutdown, the emergency generator will be started automatically.

All other systems remain in operation.

5.6.4 Level 4

Level 4 shutdowns would be initiated by external non-process factors, such as gas detection in a safe area, or a crane accident.

An example of such a shutdown is as follows:

1. Level 3 shutdown actions 1 to 3 are taken.
2. All main and emergency power generating equipment is stopped.
3. AC circuits of all uninterruptible power supplies (UPS) inverters and some DC systems are switched off. However, the following DC secure supply systems remain in operation:
   (a) explosionproof equipment required for the safe shutdown of compressors, turbines etc.;
   (b) fire and gas detection and protection systems, including the fire water pumps;
   (c) explosionproof emergency lighting, navigational aids etc.;
   (d) the shutdown system itself.

5.6.5 Level 5

A level 5 shutdown involves the shutdown of every electrical system and the isolation by explosionproof isolator of every battery, with the sole exception of the fire water pump start and control batteries. This level of shutdown is only considered during major gas leaks, blowouts etc., where there is a danger that most if not all of the installation will be covered by a gas cloud. Initiation of a level 5 shutdown is manually by an authorized operator only (usually the installation manager or his deputy). Facilities such as key override switches and timers must be provided to ensure that the system can be reinstated once the abnormal condition has cleared.

Shutdown systems are often controlled by hard-wired relay logic since, at the time of writing, operator confidence in software controlled devices has not yet reached sufficient heights for their universal adoption offshore.

5.7   Effects of ESD system on electrical systems

The examples given in Chapter 2 indicate some of the problems experienced with the interaction of the electrical and emergency shutdown systems in the isolated offshore environment.

Before completion of the system conceptual or front end design studies, it is important that:

1.   All interactions between systems are clearly understood.
2.   All the likely failure modes are known, and practical remedial measures or compensating facilities are incorporated in the design.
3.   Facilities are incorporated in the system design to ensure that once a shutdown has occurred it is possible, barring catastrophic failures, to return to production in a reasonable length of time. It should not be necessary to make temporary cable connections, or to obtain temporary equipment from the shore base, in order to do this.

It is recommended that in order to obtain the above, a failure modes, effects and criticality analysis (FMECA) be conducted at this time, possibly in conjunction with a hazop study on the process system (see Chapter 12). The output from these activities could be part of the installation safety case called for by the Cullen Report.

5.8   Uninterruptible and secure power supplies

Because of the need to maintain control of the process, and provide electrical supplies for statutory safety equipment such as navaids and marine and helicopter radios, whether generation is available or not, a number of battery fed secure supply systems are required on every offshore installation.

The grouping of these supplies so that they are fed from a common battery/charger system should be considered carefully in order to avoid common mode failures, i.e. situations where several vital supplies are lost owing to the failure of a common component. Emergency lighting fed from a central battery is particularly prone to this form of failure, as battery or distribution board failure will extinguish *all* the lamps on the system. Although luminaires with integral emergency batteries require slightly more maintenance than luminaires without batteries, the benefit, in terms of increased system reliability by avoiding the risk of common mode failure, is very considerable.

5.9 DC supplies

The majority of secure supply systems are DC, and are required for the following:

Engine start batteries The normal criterion for generator starting is that at least six consecutive start attempts each of 10 seconds cranking are possible from a fully charged battery without further recharge. In the case of fire pump start batteries, this is raised to twelve attempts from each of two banks of batteries. This would be further reinforced by a second means of starting, such as a hydraulic pump device or compressed air motor.

Class A equipment Fire and gas panels, emergency radios and public address systems are usually called category A or class A equipment. They are provided with duplicate chargers and batteries, each battery/charger being capable of supplying the rated load for the specified discharge time with mains failed. It is important to ensure that the manufacturer's design provides a fully duplicated system. There should be no common components in the rectifier or voltage regulation system, and the connection point for the two supplies, i.e. the distribution board, should be protected by blocking diodes and fuses from failure of one of the battery/charger units. It is also normal practice to obtain the AC supplies for the two chargers from different points in the supply system in order to increase overall reliability.

Class B equipment All other DC supplies, with the possible exception of some subsea equipment (see Chapter 11), are supplied by single battery/charger units. However, for improved reliability a compromise of two 100% rated chargers with one battery may be used. As most of the components of the system are in the charger, an improvement in reliability can be obtained without the extra weight or bulk of a second battery.

Switchgear tripping supplies Offshore practice is similar to that for onshore substations. For reasons of reliability, each supply should be dedicated to one switchboard. Batteries should have sufficient capacity for the tripping and closing of all circuit breakers on the switchboard in succession, twice, followed by the simultaneous tripping of all the circuit breakers. There should also be sufficient capacity to supply the continuous drain imposed by any control relays and indication lamps for a reasonable period (say 8 hours).

5.10 AC supplies

These uninterruptible power supplies (UPS) units are required where process controls, instrumentation, telemetry and telecommunications equipment require a secure source of AC power to avoid disruption of remote control, blank sections in logging records or loss of voice communication with the shore base.

Provided the platform generated frequency and voltage are reasonably stable and within a few per cent of the nominal values, the inverter will normally remain in synchronism with the generated supply. During generated voltage or frequency excursions, the inverter will isolate itself from the main supply and continue to feed the load by itself. A mains by-pass supply is also available and, if the inverter fails, this by-pass supply is automatically connected to the load. However, in some designs, if the by-pass supply is outside the inverter voltage or frequency tolerance, the by-pass supply will not be made available to the load. This can be a problem if the generated supply is being obtained from the emergency generator, which will not normally have such good voltage or frequency holding capabilities as the larger machines.

It should be remembered that the inverter itself is generating a sine wave from a power oscillator, and therefore cannot produce a fault current much greater than its rated current. For example, a 20 A inverter with the by-pass supply unavailable is able, at best, to blow a 6 A fuse in a reasonable time. If the distribution system requires larger fuses, then an inverter of higher rating will normally be required. It will not be necessary to provide a larger battery, however.

5.11 Selection of voltage tolerances

The general rule for establishing voltage tolerances is to optimize these for the minimum output cable cross-sectional area and battery size. Having been provided with or having calculated the diversified load current, discharge duration and voltage tolerances required for the equipment being supplied, the engineer can establish the minimum battery size from the manufacturer's tables.

However, this takes no account of the voltage drop in the DC output cable. This voltage drop must be catered for in the supply system to ensure that the lower tolerance of the supplied equipment is not exceeded, as shown in Table 5.1 for a nominal 24 V DC system. The fire and gas panel supply shown in the table requires a 150 mm^2 cable because the voltage drop has been restricted to 0.5 V DC. If the minimum voltage at the charger is raised to 23.1 V DC, it would be possible to reduce the cable size to 70 mm^2. However, in order to raise this voltage, it would be necessary to increase the battery capacity, assuming the required discharge time remains constant.

5.12   Batteries

5.12.1 Types of battery

There are three types of battery used commonly offshore, as follows.

Table 5.1

Equipment title	Load A	Cable (mm²/length)	Volts at charger		Volts at equipment	
			min.	max.	min.	max.
Fire/gas panel	40	150/50 m	22.5	26.4	22.0	26.4
ESD panel	12.5	35/45 m	23.1	27.0	22.5	26.9
PA system	25	95/100 m	22.6	26.4	21.6	26.4

Conductor resistances obtained from BICC cable selector S22

5.12.1.1 Lead acid Planté cells
This type of cell is appropriate for most applications where a suitable ventilated battery room can be provided. It should not be used in a hot vibrating environment such as adjacent to engines. This is because the electrolyte will tend to evaporate and require frequent topping up owing to the heat, and the vibration will tend to cause the oxide coating to fall off the plates. With careful maintenance these batteries may last in excess of 20 years.

5.12.1.2 Lead acid recombination cells
These cells are a relatively recent development and provide several advantages over Planté cells. The cell contains no free electrolyte, all the electrolyte being contained in an absorbent blotting-paper-like material. Except for a small safety vent, the cells are sealed for life; unlike Planté cells, they normally emit only molecular quantities of hydrogen. Because of their recent availability, no operational life expectancy figure is available, but it is known from present experience to be in excess of eight years.

Excessive charging of recombination cells will permanently damage them by electrolysing the small amount of electrolyte they contain. Once this has been vented, it cannot be replaced. Therefore it is vital that, with the chargers used, there is a very low probability that a fault will occur which allows charging voltages greater than the maximum recommended by the manufacturer. A boost charging facility must not be fitted for the same reason. Also for the same reason, engine starting batteries of this type must not be connected to an engine driven charging alternator, as this will more than likely exceed the permitted charging voltage.

If, on discharge, the cell voltage falls below a certain minimum value, it will no longer be possible to recharge the cells affected, making it necessary to replace them. Therefore it is important that during the installation and commissioning period, cells are fully recharged at least every six months.

Despite the above considerations, these batteries are becoming very popular, mainly because of the saving that can be obtained from the reduction in maintenance required, and since reliable electronic chargers are available with good voltage regulation and current limiting facilities.

5.12.1.3 Nickel cadmium cells
These cells are generally more able to withstand heat, shock and vibration than both of the lead acid types. A nickel cadmium battery is 35% lighter than the equivalent lead acid one. However, nickel cadmium batteries are considerably more expensive, especially in comparison with recombination

cell batteries of the same capacity and voltage. Although able to endure longer periods of inaction without charge, they may require more skilled maintenance than lead acid batteries for the following reasons:

1.   There is a memory effect associated with repetitive cycling of the battery which may cause the battery voltage to be depressed below the value expected from its discharge characteristic. The effect is usually temporary and can be reversed by reconditioning cycles.
2.   Sustained boost charging may have the effect of depressing the discharge voltage slightly. As above, this can be reversed by reconditioning cycles.

5.12.2 Boost charging facilities

It is common to find a boost charge mode available on battery chargers, the intention being to enable the operator to reduce the battery recharge time. With the electronic chargers now available, battery recharge durations are short enough for most purposes without recourse to a manually initiated boost charge mode. Boost charge facilities are not recommended on any battery systems because of the excessive gassing they can cause. If they must be fitted, some form of timer should be incorporated in order to prevent sustained overcharging. They should not be fitted to the chargers of sealed lead acid cells if permanent cell damage is to be avoided.

5.12.3 Ventilation and housing of vented batteries

Special provisions must be made for the housing of batteries offshore in order that the risks due to electrical ignition hazard, evolved hydrogen and electrolyte spillage are minimized.

The battery room provided should be lined on walls and floor with a rubber based or similar electrolyte (either sulphuric acid or potassium hydroxide) resistant material. Owing to the violence of any contact between the two forms of electrolyte, it is not advisable to house a mixture of lead acid and nickel cadmium cells in the same room. Inadvertent topping up with the wrong type of electrolyte would also be extremely hazardous for the personnel concerned.

Both types of cell evolve hydrogen, and this must not be allowed to build up to explosive levels. A ventilation system must be provided sufficient to dilute the hydrogen well below this level at all battery states. A hydrogen detector is usually provided, and any increase in hydrogen concentration would be indicated at a manned control room for the installation. Cells should be racked in tiers so that each cell is easily accessible for maintenance or replacement.

To cater for major gas leak conditions, an explosionproof isolator must be provided for each battery. These double-pole isolators would be remotely operated through the ESD system should such a serious gas leak occur.

5.12.4 Sealed cells

Although very little hydrogen is emitted from lead acid recombination cells, it is still advisable for batteries of significant size to be located in a dedicated purpose-built room. Failure of the battery charger to limit charge voltage would lead to a short-lived evolution of hydrogen, and the room ventilation would need to cater for this. A dedicated room also prevents the access of unauthorized personnel who may not be aware of the dangers associated with stored electrochemical energy; tools or other metal objects inadvertently placed across battery terminals will cause heavy short-circuit currents to flow.

Chapter 6

Generation and distribution switchgear and transformers

6.1 General requirements

Switchboards are the nodal points in any electrical system. They must be designed to provide a safe, reliable means of carrying and directing current flows where they are required, and must be able to withstand the thermal and magnetic stresses involved in clearing faults. This fault capability must include:

1. With the circuit breaker initially connecting two healthy sections of the system, the ability to interrupt a fault immediately after it occurs in a fast and reliable manner. This is known as the fault breaking capability.
2. The ability to connect a live section of the system to a faulty section and then to immediately interrupt the resulting fault, in a fast and reliable manner. This is known as the fault making capability.

For an in-depth discussion on the operation of circuit breakers, the reader is recommended to refer to *The J&P Switchgear Book*. The following is a brief introduction to the principles, and should assist the engineer in selecting switchgear of adequate performance for the application.

6.2 The mechanism of short-circuit current interruption

Figure 6.1 (a) shows the DC or asymmetrical component that will exist in the current flowing in at least one of the phases unless the short-circuit appears when the circuit voltage is at a maximum. This voltage decays after a few cycles and the resulting decrement curve can be seen. On an offshore platform, where power is more than likely to be locally generated, there will also be an AC or symmetrical decrement associated with the dynamic reactances of the generators. A typical generator decrement curve is shown in Figure 6.2.

Fault-making and breaking currents for particular operating configurations must be calculated in a standard manner if the performance of switchgear from different manufacturers is to be compared with any degree of consistency.

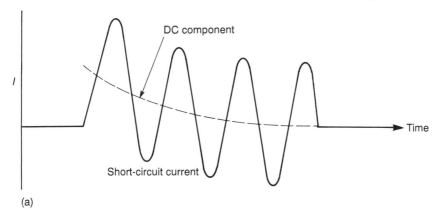

(a)

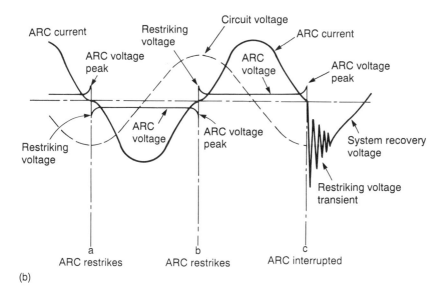

(b)

Figure 6.1 Interruption of asymmetrical short-circuit current: (a) short-circuit current (b) circuit breaker current interruption

Figure 6.3 shows waveforms for two fault situations: (a) at system voltage peak, and (b) at system voltage zero. Situation (a) shows no DC component and the peak current is $\sqrt{2}$ times the symmetrical current. In situation (b), the maximum DC component is present and the peak current is $2\sqrt{2}$ times the symmetrical current. The power factor of the fault current depends on the R/X ratio of the circuit, where R is the effective resistance of the circuit and X is the effective reactance. If the total impedance of the circuit was only resistive, the fault current which flows would be symmetrical, irrespective of the point on the current wave at which the fault occurs. Therefore the asymmetry and hence the ratio of RMS symmetrical

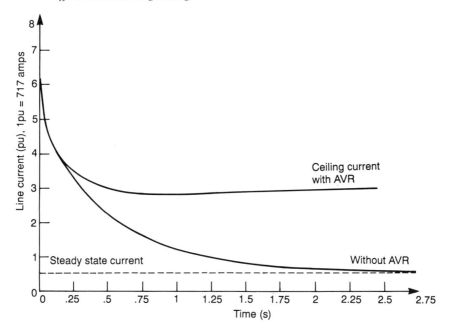

Figure 6.2 Typical generator decrement curve: $I_{SC}=I_{FU}/X''_d$, where X''_d is generator subtransient reactance

current to peak current vary with power factor. This is shown by Figure 6.4.

Three British and European standards give methods of fault current calculation and each differs from the other to some extent, as explained in Table 6.1. The BS 3659 method of calculation is the simplest. It will result in a pessimistic value, and will possibly lead to the selection of switchgear which is overrated for the faults it will be required to handle. Each of the methods in the table is successively more involved, with IEC 363 being the most involved and hence the closest to 'reality'. It is questionable whether the pessimistic value obtained can be used as a safety margin since, without doing further calculation, the extent of the safety margin will be unknown. However, the simpler methods do provide a good way of checking that existing switchgear is able to cope with new values of fault current after a system modification or expansion. Examples 6.1 and 6.2 show typical calculations for transformer and generator fault currents respectively. For a second method of calculating motor fault current contributions, refer to Cooper, McLean and Williams (1969).

Example 6.1
Figure 6.5 shows the trivial scheme used in this and the next example.

This example deals with the transformer circuit, shown in Figure 6.6. All impedances are to a 100 MVA base. The cable impedances are based on the BICC cable selector S22.

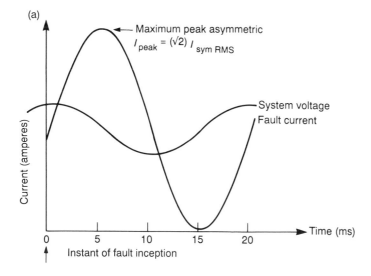

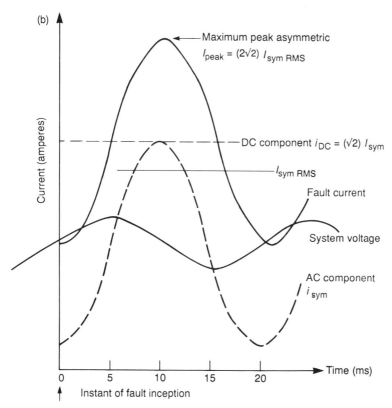

Figure 6.3 Fault inception at different points on the waveform: (a) system voltage peak (b) system voltage zero

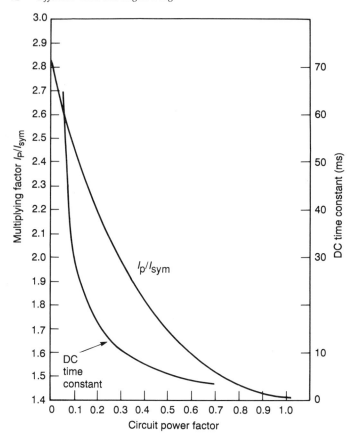

Figure 6.4 Graph of power factor against I_p/I_{sym} and DC time constant: no AC decrement

Calculation of breaking current

This method is applicable to BS 3659, BS 4752 and IEC 363 (no AC decrement).

$$\text{total impedance} = 2.193 + j\,8.12$$
$$|Z| = 8.4$$
$$\text{power factor} = 0.26$$
$$\text{MVA} = \frac{100}{8.4} = 11.9$$

Therefore

$$I_{sym}\,(t=0) = \frac{11.9}{(\sqrt{3}\,) \times 0.415} = 16.56\,\text{kA}$$

Calculation of peak current for CB making duty
BS 3659: fixed DC decrement

$$I_p = 2.55\,I_{sym} = 42.23\,\text{kA}$$

Table 6.1

Type of circuit	Standard specification	Method of calculation	
		Breaking current	Making current
Transformer	BS 3659	No AC or DC decrement	Fixed DC decrement
	BS 4752	No AC or DC decrement	DC decrement based on Figure 6.4
	IEC 363	No AC or DC decrement	DC and AC decrements calculated
Generator	BS 3659	No AC or DC decrement	Fixed DC decrement
	BS 4752	No AC or DC decrement	DC decrement based on Figure 6.4
	IEC 363	AC decrement based on generator transient impedance	DC and AC decrements calculated
Motor contribution	BS 3659 and BS 4752	Not covered	Not covered
	IEC 363	Approximation based on $I_{sym} \approx 4 \times$ full load current	Approximation based on $I_p \approx 8 \times$ full load current

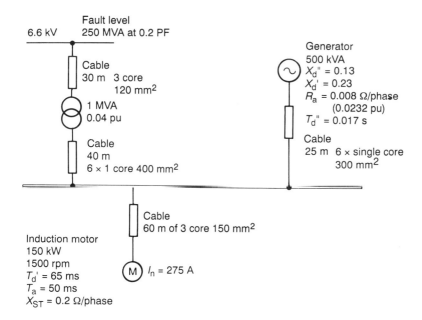

Figure 6.5 Trivial scheme for example calculation of switchgear breaking and making current duty

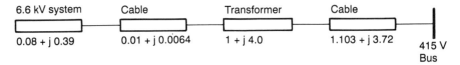

Figure 6.6 Transformer circuit for Example 6.1

BS 4752: DC decrement
The power factor is 0.26. Therefore, from Figure 6.4, I_p/I_{sym} = 2.06.
Hence

$$I_p = 2.06 \times 16.56 = 34.11 \, \text{kA}$$

IEC 363: DC and AC decrements
The I_{DC} at fault inception is $(\sqrt{2}) \, I_{sym}$ = 23.42 kA. From Figure 6.4, the DC
time constant at a power factor of 0.26 is 12 ms. Thus

$$I_{DC} \, (t = 10) = 23.42\text{e}^{-10/12} = 10.18 \, \text{kA}$$

Now I_{sym} at 10 ms = I_{sym} at fault inception (no AC decrement from
transformer supply).

So $I_p = (\sqrt{2}) \, I_{sym} + I_{DC} = 23.42 + 10.18 = 33.6 \, \text{kA}$

Example 6.2
This example deals with the generator circuit, shown in Figure 6.7. All
impedances are to a 100 MVA base. The cable impedances are based on
the BICC cable selector S22.

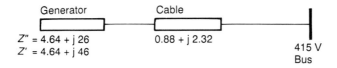

Figure 6.7 Generator circuit for Example 6.2

Calculation of breaking current
BS 3659 and BS 4752: *no AC decrement*

total subtransient impedance = 5.52 + j 28.32

$$|Z| = 28.85$$
$$\text{power factor} = 0.19$$
$$\text{MVA} = \frac{100}{28.85} = 3.47$$

Therefore

$$I'' = \frac{3.47}{(\sqrt{3}) \times 0.415} = 4.82 \, \text{kA}$$

IEC 363: AC decrement
  total transient impedance $= 5.52 + j\,48.32$
$$|Z| = 48.63$$
  power factor $= 0.11$
$$\text{MVA} = \frac{100}{48.63} = 2.06$$

$$I' = \frac{2.06}{(\sqrt{3}) \times 0.415} = 2.86\,\text{kA}$$

$T_d'' = 17\,\text{ms}$

Therefore

$$I_{sym}(t = 10) = (4.82 - 2.86)e^{-10/17} + 2.86$$
$$= 1.088 + 2.86 \quad \text{(from no load)} = 3.95\,\text{kA}$$

Allowing for generator preload,

$$I_{sym}(t = 10) = 3.95 \times 1.1 = 4.34\,\text{kA}$$

Calculation of peak current for CB making duty
BS 3659: fixed DC decrement

$$I_p = 2.55\,I'' = 12.3\,\text{kA}$$

BS 4752: DC decrement
The power factor is 0.19. Therefore, from Figure 6.4, $I_p/I_{sym} = 2.21$. Hence

$$I_p = 4.82 \times 2.21 = 10.65\,\text{kA}$$

IEC 363: DC and AC decrements
The I_{DC} at fault inception is $(\sqrt{2})\,I'' = 6.82\,\text{kA}$. From Figure 6.4, the DC time constant at 0.19 power factor is 16.5 ms. Thus

$$I_{DC}(t=10) = 6.82e^{-10/16.5} = 3.72\,\text{kA}$$

$$I_p = (\sqrt{2}) \times 3.95 + 3.72 = 9.3\,\text{kA}$$

The same methods may be adopted for calculating the short circuit currents associated with the motor circuit. (Refer to table 6.1 for values of I_{sym} and I_p)

6.3 Types of interrupter

6.3.1 HRC cartridge fuses

A fuse is basically a device with a central conductor that is designed to melt under fault conditions. Interruption is achieved by spacing the two ends sufficiently far apart for the arc to be naturally extinguished. However, to obtain a consistent performance characteristic, fuse elements are carefully designed for particular voltage and current ratings. The ceramic/cartridge tube is filled with powdered quartz and sealed. The silver element is not a continuous strip of silver, but is necked in short sections to reduce pre-arcing time. It may also have sections of low melting point, M-effect material to improve performance at low fault levels (see Figure 6.8). The physical design of the element, such as the length and the shape of the necks, will depend on the application of the fuse and its operating voltage. If the maximum arc voltage is exceeded, the pre-arcing time will be reduced and the fuse will not operate according to its standard characteristic.

Since a fuse is purely a means of protection against overload and fault currents, isolators or contactors are installed in series to carry out normal switching operations.

The advantage in using fuses is that the resulting switch-fuse device is often less expensive, smaller and lighter than the equivalent circuit breaker, particularly in the higher power ranges. Exceptions to this for low-voltage distribution equipment are some of the higher-performance current limiting miniature circuit breakers. These will be discussed later in the chapter.

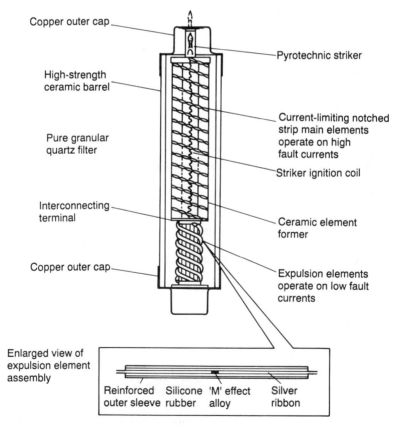

Figure 6.8 Diagram showing the interior of typical HRC fuse. (Courtesy of GEC Alsthom Installation Equipment Ltd.)

The disadvantages with fusegear are twofold. First, HRC fuses must be replaced once operated. No matter how reliable the circuit is, some holding of spare fuses will be necessary, and replacement of the larger bolted fuses is time consuming. Permanent power fuses are available, however, which use the thermodynamic characteristics of liquid sodium to interrupt fault currents for a few milliseconds, during which time a suitable switching device is operated to isolate the faulty circuit. This type of fuse is not (to the author's knowledge) used offshore. Secondly, the standard ranges of fuses available are limited in current and fault capacity below the maximum ratings available in switchgear ranges.

6.3.2 Air circuit breakers

An air circuit breaker (ACB) is a device where the circuit is made or interrupted by moving contacts located in atmospheric air. The ACB relies on moving the contacts sufficiently far apart to extinguish the arc under short-circuit conditions. To assist in this operation, the design of ACBs has been greatly improved since the original invention, notably in the following ways:

1. 'Trip-free' mechanisms are used, in which the geometry of the mechanism is such that tripping can occur even while a closing operation is under way.
2. Separate arc and current carrying contacts give better thermal rating.
3. Blowout coils and arc chutes are employed. A blowout coil uses the fault current to produce a strong magnetic field which pulls the arc away from the contacts into the arc chute. The arc chute is a series of parallel insulators designed to extinguish the arc by splitting and cooling it. Because of the need for blowout coils, performance will be partially related to the fault current magnitude. This effect will appear as a critical current below which the arc is not drawn into the chute; arc contact wear will be accelerated, and the risk that fault clearance will not be achieved will be increased.

This type of switchgear also tends to be bulky, mechanically complex and hence more costly compared with other forms now available. However, this equipment is well proven and its use well established offshore, since it has the ability to interrupt the high magnitudes of fault current found in offshore systems.

6.3.3 Bulk oil circuit breakers

The bulk oil circuit breaker is a device in which the moving contacts are totally immersed in a container of mineral oil. As the oil is a better insulant and has a higher specific heat than air, contact gaps may be reduced, and the better heat dissipation means that the overall cubicle size may be reduced compared with those for an ACB of the same rating. Arc extinguishing is assisted by convection currents in the oil, produced by the heating effects of the arc. No weight saving is likely, however, because of the weight of oil required. Arcing within the oil causes hydrogen gassing which must be vented from the container; although with modern designs the risk is minimal, open flames or even explosions may occur under severe fault clearing duty or normal fault clearing with badly contaminated oil. The insulating oil can deteriorate over a number of operations or after a long period, and requires regular sampling and testing. Sufficient replacement oil must be stocked offshore or at least made available offshore prior to testing.

Although this form of circuit breaker is still in common use onshore and has been installed for main switchboards offshore, it is now generally considered undesirable in an offshore environment and its use may be questioned by a number of certifying authorities and underwriters.

6.3.4 Limited oil volume circuit breakers

If oil is injected at high velocity between the contacts as they open, an efficient means of arc extinguishing may be obtained. The amount of oil required is much less than with a bulk oil circuit breaker, and therefore the fire risk is reduced. However, as with the air and bulk oil breakers, maintenance requirements are usually heavy compared with vacuum and sulphur hexafluoride types (see following).

6.3.5 Vacuum circuit breakers and contactors

At the heart of the vacuum circuit breaker is a device called the vacuum interrupter, one of which is required for each pole. The interrupter (see Figure 6.9) consists of a ceramic tube with metal seals at both ends. The fixed contact is mounted on one metal seal, while the moving contact at the other end of the tube is free to move inside a metal bellows which maintains the vacuum seal. The ceramic tube is usually in two parts, to allow the insertion of a sputter shield designed to prevent contact metal condensing on the ceramic tube and providing a conducting path between poles. On modern interrupters the sputter shield is now an integral part of the contact assembly, as can be seen in Figure 6.9. The vacuum is much harder, at 10^{-8} to 10^{-5} mbar, than in a fluorescent lamp, and should only allow a few free molecules. Thus the majority of ionized particles required to support an arc are provided by molecules from contact metal, the metallurgy of which is vital to the satisfactory operation of the interrupter. Hard contact metals will not provide sufficient molecules and arcing will be extinguished prematurely, leading to current chopping and high-voltage transients. However, if the contact metal is too soft, contact wear will be accelerated and contact welding may occur. The difference in contact metal is one of the essential differences between the interrupter and the contactor bottle, the contactor having a softer contact metal. Vacuum contactors still have a significant fault rating, of around 7 kA, but, as with other forms of contactor, require to be protected by fuses for currents of greater magnitude if contact welding is to be avoided. The circuit breaker resulting from this technology is compact and light, with a much reduced maintenance requirement. Foundation shock due to operation is very much reduced, owing to the lightness of contacts and the small distances they have to move apart for fault clearance.

Interrupters and contactors can be checked for adequate vacuum by applying a test voltage across the contact when in the open position. A

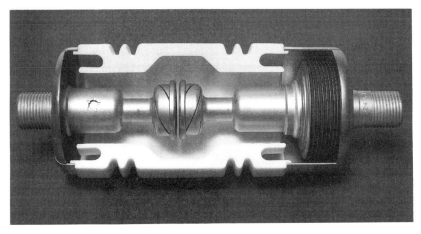

(a)

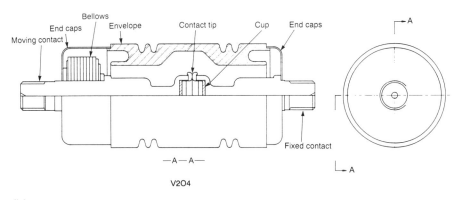

(b)

Figure 6.9 Interior of typical vacuum interrupter. (a) Photograph of sectioned interrupter; (b) sectional drawing. (Courtesy of GEC Alsthom Vacuum Equipment Ltd.)

potential of 25 kV AC is required for testing a 12 kV interrupter. Owing to the permeability of the various ceramic and metal construction materials, the operational life is limited because of a slow loss of vacuum, but by using a getter it is normally in excess of 20 years.

Short-circuit ratings of up to about 40 kA are available for circuit breakers operating up to 12 kV. This may not be sufficient on some of the larger installations, where prospective fault currents and operating voltages may be higher.

Because of the size, weight and space advantages of vacuum interrupters and contactors, and the reduced fire hazard, this type of equipment is recommended for offshore use.

6.3.6 Sulphur hexafluoride (SF$_6$) circuit breakers

The SF$_6$ circuit breaker competes with the vacuum device for low weight and bulk and can be considered as an alternative in most cases. Sulphur hexafluoride has a dielectric strength several times that of air, and good heat dissipation properties. Low pressures are required to be maintained (usually less than 2 bar) and it is unusual if topping up is required more than once every two years. Suitably high fault ratings are available for most offshore situations.

6.4   Switchboard construction

The following paragraphs list some important recommendations for switch-gear procurement specifications. The configuration of switchboards is discussed in Chapter 2, and reliability aspects in Chapter 12.

In all cases, offshore switchboards must be highly resistant to the salt-laden corrosive atmosphere, and sufficiently moisture protecting to prevent the ingress of water, particularly from above (IP54 minimum). The enclosures must also physically protect operators from any arcing, flames or flying fragments due to maloperation.

Major changes to switchboards, once they are installed offshore, are extremely expensive. Depending on the planned operational life expectancy of the offshore installation, as much spare equipped capacity should be incorporated into the switchboard as possible. The spare auxiliary contacts of circuit breakers, contactors, switches and relays should all be wired out to terminals as a matter of course by the manufacturer. Whether racked or single mounting methods are used for protection relays, space for additional relays or spare relay rack spaces should be provided. Facilities for adding extra switchboard cubicles at either end should also be provided.

It is also important that facilities for cabling are as flexible as possible, and that sufficient space is available for installation of the largest cables without exceeding bending radii limits. Neutral connection arrangements for outgoing supplies with neutrals must be provided. Facilities need to be provided for the earthing of cable armour. If cable entry is at the top, there should be no risk of moisture entering via the cable entry even if the cables do not have drip loops.

The space requirements for circuit breaker handling trucks must not be forgotten.

Functional test facilities should be built into the switchboard using a test panel cubicle with bus wiring to each unit to avoid dangerous trailing leads. The test facility must be interlocked to ensure that live operation of field equipment is impossible using the test panel. Every unit on the switchboard will require suitable permanent labelling and interlocking, shuttering and maintenance padlocking facilities. In fact, all the usual requirements for onshore substations such as safety rubber matting, earthing facilities and safety testing equipment will be required.

Protection relay schemes are covered in Chapter 9.

6.4.1 Main switchboards 6.6 to 13.8 kV

The overriding consideration for any main switchboard must be the short-circuit capability of the circuit breakers, because of the fault current capability and proximity of the installed generation. Generator operational configurations which produce prospective fault MVAs of more than 1000 should be avoided, as downstream equipment and cables would require to be of special non-standard manufacture, with all the expensive development and testing this would entail, in order to obtain sufficient rating.

Table 6.2 shows typical examples of switchgear available in the early 1990s. As discussed earlier in the chapter, a check will have to be made to ensure that the asymmetrical breaking capacity is adequate, allowing for the decrement in the value of this between fault inception and contact opening. Generator AC decrement must also be taken into account.

Table 6.2

Manufacturer	Type		Fault capacity (MVA)	Max. operating voltage (kV)
Reyrolle	Class SA	(air break)	930	13.8+
GE (USA)	Power Vac	(vacuum)	1000	13.8+
Whipp & Bourne	DV40	(vacuum)	830	12.0+
Merlin Gerin	FC3	(SF$_6$)	950	13.8+

Providing the switchgear fault make rating is adequate, problems with fault break ratings may be overcome by delaying the circuit breaker opening until the fault current has decayed to a value within the rating. The use of busbar reactors is not recommended owing to offshore weight and space limitations. Because of the high prospective fault currents, it is likely that any large motors supplied directly from this switchboard will require circuit breaker rather than fused contactor switching.

To avoid shutting down generators or other vital equipment to carry out maintenance on the switchboard, a duplicate busbar switchboard may be considered. This is not often specified, however, because of the extra complexity, cost, weight and space involved.

If it is likely that further generators will be required to be installed, owing to a later operational phase such as artificial lift, then the switchboard will require to be rated for this future load and fault rated for the future prospective fault current capacity of the expanded system. Sufficient spare equipped circuit breakers should be provided for the expansion.

6.4.2 Large-drive switchboards 3.3 to 6.6 kV

Development of motor control gear at up to 6.6 kV has resulted in very compact units where relatively low load currents are switched by vacuum contactors protected from short-circuit faults by suitable HRC fuses. For incoming and outgoing distribution, circuit breaker cubicles are provided, the whole forming a composite switchboard of low weight and compact dimensions.

The prospective fault level on this switchboard can be regulated to some extent by adjusting the reactance of the supply transformer windings. Therefore motor control is usually by fused contactor rather than circuit breaker.

6.4.3 Utility services and production switchboards

Because of the interdependence of various systems on an offshore installation, as can be seen by the examples in Chapter 2, the low-voltage switchboards must be considered just as vital as their medium-voltage neighbours.

Maintenance of circuits for such supplies as machinery auxiliaries and hazardous area ventilation must be given high priority. An example of a generator lube oil auxiliary system is given in Chapter 3.

6.4.4 Emergency switchboards

The function of the emergency switchboard is described in Chapter 1.

It is beneficial to provide synchronizing facilities for the switchboard's associated emergency generator. The generator has automatic start facilities which will initiate a start following a main generation failure, provided the start signal is not inhibited by one of the safety systems. The synchronizing facility gives a convenient means of routine load testing for the generator, and allows for changing over to main generation after a shutdown incident, without a break in the supply.

The switchboard should also include facilities to prevent the generator from starting when a fault exists on the switchboard. Interlocking must be provided between the emergency generator incomer and the incomer from the rest of the platform power system if synchronizing facilities are not available.

As the emergency switchboard usually feeds all the AC and DC secure supply battery chargers and other vital equipment, it is important that planned switchboard maintenance outages are catered for in the design. It is not usual to go to the expense of a duplicate bus switchboard, but certain battery chargers and other vital equipments are usually fed from an alternative switchboard via a changeover switch. These supplies should also include those necessary for starting other generators and for safe area ventilation, the basic philosophy being to allow continued safe oil production whilst the switchboard is being serviced.

6.5 Drilling supplies

The drilling electrical system is usually independent of the installation system, with its own diesel generation. The reason for this is partly to do with organization, since the drilling may be carried out by a different company which provides the complete drilling package, including generation and switchgear.

If the drilling system is of the same operating frequency as the rest of the platform, then an interconnector of some kind between the systems is mutually beneficial, provided there is no equipment in the main system that is particularly sensitive to the harmonics generated by the silicon controlled rectifier (SCR) equipment. The reliability and maintainability of the drilling electrical system is vital, as failures at particular times in the drilling operation may increase the risk of blowouts or cause the abandonment of a producing well.

The switchgear used for drilling distribution is of the conventional motor control centre type, with the exception of the SCR cubicles. The SCR cubicles cannot be isolated individually owing to the permanent interconnection arrangement between each of the variable speed drives. An assignment switch on the driller's console allows the connection of any SCR cubicle to any DC drive motor in the system (see Figure 6.10(a)). This arrangement allows for DC drive motors to be reassigned to another SCR cubicle if a fault develops in the first cubicle. To allow plenty of ventilation, the SCR cubicles are of a much more open design than other offshore switchboards. The assignment contactors are usually arranged in the upper section of the panel, while the SCR assemblies are in the lower section. Figure 6.10(b) shows a typical SCR cubicle schematic diagram.

To avoid obstruction of cooling air, there are usually no insulating barriers between the interconnecting busbars, contactors and equipment within the cubicles. However, it is sometimes necessary, if drilling is to continue, for the rig electrician to change SCRs with the cubicle only isolated by the assignment contactor at the top. This problem is usually overcome by using a removable insulating barrier which can be carefully located below the contactor before working on the SCR assembly.

6.6   Living quarters supplies

Electrical supplies for living accommodation are important for the well-being of the offshore staff. Failure will not only bring discomfort, but will present a serious health hazard, when frozen food begins to thaw, toilets cannot be flushed etc.

As mentioned in Chapter 2, domestic pipework should be kept out of switchrooms where possible. Where pipes do have to pass through switchrooms, possible sources of leaks such as flanges, valves etc. should be kept well away from switchboards or avoided altogether.

Most offshore accommodation is sealed and positively pressurized to prevent ingress of gas should any process leaks occur, and so the operation of air conditioning is also vital.

Should some serious incident result in a gas cloud developing, leading to ingestion of gas by the accommodation module HVAC, all generation is usually shut down and doors and fire dampers are closed as automatic actions of the ESD and fire and gas systems. The actions taken, however, will depend on many design and operational factors associated with the particular installation. When the incident is over, certified hazardous area fans fed from another source may be used to purge the remaining gas from the module. Some operators prefer to allow natural ventilation to clear the gas, as purging fan systems have been known to ignite the gas and cause an explosion.

84

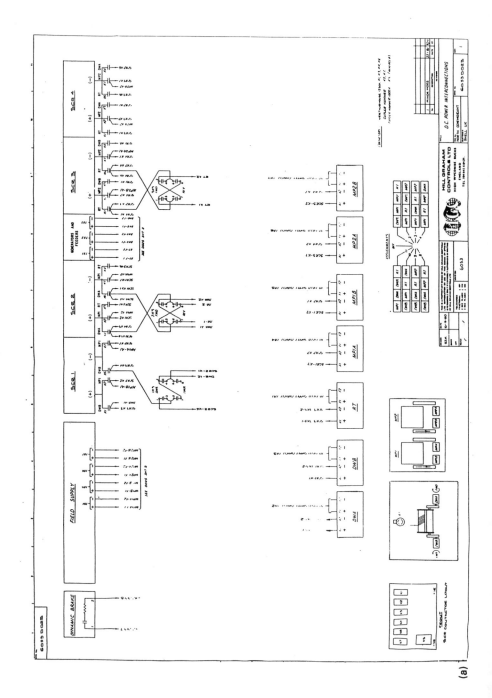

(a)

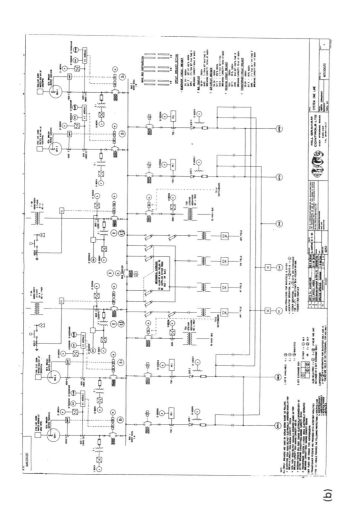

Figure 6.10 Schematic diagram of typical drilling SCR cubicle. (Courtesy Hill Graham Controls Ltd.)

(b)

A serious fire outside the accommodation module may lead to its envelopment in dense smoke. Smoke detectors located at the ventilation air intakes should signal the fire and gas systems to close down the ventilation system automatically, shutting fire dampers and doors to seal the module.

6.7 Process area distribution

By far the largest power consumers on any offshore oil installation are the process modules. Apart from the large process drives, power is required for level control and circulating pumps, agitators, centrifuges, compressors and ventilation fans. Lighting and instrumentation power must be provided, and also sockets for welding transformers and other temporary equipment for maintenance. The production switchboards located in safe area switchrooms provide the majority of control gear for this smaller type of equipment. Welding sockets and portable tool sockets are equipped with isolators. A popular type of socket unit is certified Ex'e', but an explosionproof cassette containing the isolator contacts is housed within the Ex'e' enclosure. All such sockets are fed from shutdown contactor feeders in the production switchboards, so that should gas be detected, all portable equipment in the area may be immediately isolated.

Lighting distribution is discussed in Chapter 10.

6.8 Transformers

Offshore distribution transformers are usually of either sealed silicon oil filled or encapsulated resin types. Standard mineral oil filled types are too great a fire hazard, and askarel based insulants are regarded as a health hazard owing to the presence of PCBs. Air cored types are not recommended offshore because the salt-laden environment tends to lead to insulation problems. The sealed silicon filled type has the advantages that a faster repair can normally be obtained, and Buchholz pressure sensing for winding faults can be fitted. Being heavy devices, transformers need to be checked against switchroom maximum floor loadings.

If a transformer is used to interconnect the main system with the drilling system, the heating effect of SCR harmonics may need to be considered.

Chapter 7

Cabling systems and equipment installation

When a new offshore design and construction project gets under way, there is obviously a great deal of organization involved. A good, disciplined organization structure with clear, practical office procedures assists the engineers involved in producing the design, selecting the materials and equipment, and supervising the construction. A detailed examination of design and construction office technical documentation is beyond the scope of this book, but a brief discussion is provided in the first section for those not yet familiar with the subject.

7.1 Technical organization of the design programme

7.1.1 Project procedures

The end product of any design office is to produce a set of documents which perform the following functions:

(a) a design document which fully describes the design of the new system, and in which each stage is adequately supported by logical reasoning, calculations and diagrams, sketches etc.;
(b) a material list which identifies every material and equipment component, down to the level of cable clips, nuts, bolts and washers if need be;
(c) an installation workscope document which fully describes, in text, drawings, schedules and diagrams, how and in what sequence the equipment and material are to be installed and commissioned.

This end product is not arrived at in one long session, but is broken up in a series of submissions or packages. The level of detail will also increase as the project progresses. Typically, this would occur as follows:

1. Conceptual or front end engineering study. This document should put forward technical and economic arguments for and against the feasibility of various alternative design schemes, with recommendations as to which approach should be adopted.
2. A detailed cost estimate, aiming for an accuracy of say +20% and −10% of the actual project cost. At this stage, a planning network for

the whole project and an installation procedure will need to be worked up to a significant degree of detail to enable the required level of estimate accuracy to be achieved.

3.   When the skeleton scheme for the accomplishment of the whole project has been produced, detail work on the project proper can then commence. The final documents produced may be loosely divided into two categories: those required for construction, which contain material lists and commissioning and installation procedures; and those required for technical approval, which contain detailed descriptions and design calculations for each item of equipment, including information on its intended location on the offshore installation. These will be sent to a certifying authority such as Lloyd's Register or Det Norske Veritas for approval.

4.   Whilst the preparation of design packages is under way, procurement specifications for all large items of equipment must be prepared. It would be of great assistance to the engineers preparing the specifications if this particular task could be delayed until the associated design packages have reached an advanced state of preparation. Unfortunately this is invariably not the case, because manufacturers' delivery times for generators, large motors and switchgear are usually in excess of six months, making it imperative that orders are placed early in order to adhere to the project programme. Changes to switchgear to cater for changes in motor and distribution equipment ratings are therefore unavoidable, and may reach cost figures of the same order of magnitude as the original switchgear order. It is hoped that project accountants bear this in mind!

Note that the above work is carried out in the normal commercial environment, with tendering and award of contracts for the various stages by the oil company to various design contractors.

7.1.2 Office environment and procedures

The project working conditions should be such that the people involved are assisted in their efforts to ensure that the end product design functions correctly, is as safe as practicably possible, has the least environmental impact and is cost effective. In my experience, both the design team's physical environment and the administrative and quality assurance procedures adopted will significantly affect the degree of excellence of the end product.

7.1.2.1 Office working environment
Many books, both serious and humorous, have been written on this subject, but the principal points are:

1.   Provision of good heating, ventilation and lighting.
2.   Sufficient work and storage space for each person, and the maintenance of uncluttered accessways throughout the working areas. The premises chosen to house the project should be large enough to cater for the project team at the peak manning level, otherwise the team

will be broken up into Portacabins etc. at the most critical period of the project.

3. Good sound insulation from outside noise, and the segregation or soundproofing of noisy office machines such as photocopiers and printers.
4. Adequate provision of telephones, computers, software, drawing boards, catalogues, standards, codes of practice etc.

7.1.2.2 Office procedures

This subject should be dealt with in detail by the quality assurance (QA) manual of the company concerned. The manual should be based on the guidance given in the BSI Handbook 22, and in particular the quality assurance standard BS 5750. The general principles are as follows:

1. The project organization should appoint a QA experienced management representative whose sole responsibility is to implement and maintain a satisfactory QA system. The representative must be given the necessary authority to carry out his or her duties, otherwise the quality procedures manual will not be taken seriously by the design team, especially at the beginning of the project when the need for strict procedures is not obvious.
2. Every design document must be signed as checked and approved by those officially authorized to do so.
3. The circulation of all project documentation must be such that those whose responsibilities are affected by a particular document receive that document in good time to take any necessary actions. This can be difficult to accomplish if the document has to be commented on by a series of people within a very limited period.
4. A quality control system is necessary in order that all items of equipment are inspected at the manufacturer's works and appropriate tests witnessed to ensure that the equipment is fit for its purpose before it is delivered and installed.

7.1.3 Drawing representation

The following types of electrical drawing are required for most offshore design applications.

7.1.3.1 Single-line diagram

This diagram is usually the prerequisite for any electrical system design. It will be developed through most of the design period, starting as a simple sketch and finally showing details of type and rating for circuit breakers, transformers, contactors, busbars, cables, protection and control relays, metering, interlocking and other safety devices. Indication and control circuitry will also be shown in abbreviated form, although identical circuits will normally only be shown once, with appropriate references. A schedule may also be shown so that numerical information of a repetitive nature can be listed separately. Part of a typical fully developed single-line diagram is shown in Figure 7.1.

E / ITR / C	STANDBY EARTH FAULT INVERSE TIME RELAY (CURRENT OPERATED).
OE / UE / R	OVER EXCITATION, UNDER EXCITATION RELAY.
3 / OCR / ITR	3 POLE IDMT OVERCURRENT RELAY.
2OC / ITR / E	2 POLE OVERCURRENT INVERSE TIME & 1 POLE EARTH FAULT RELAY.
3HS / 2OC / ITR / E	3 POLE HIGH SET OVERCURRENT (INSTANTANEOUS) AND 2 POLE IDMT OVERCURRENT & 1 POLE INSTANTANEOUS EARTH FAULT RELAY
3HS / OC / ITR / E	3 POLE HIGH SET OVERCURRENT RELAY (INSTANTANEOUS) & 1 POLE INSTANTANEOUS EARTH FAULT RELAY.

EC CIRCUIT EARTHING FACILITIES   EB BUSBAR EARTHING FACILITIES

OSP / R   ELECTRICAL & MECHANICAL OVERSPEED RELAY.

DFR   DIODE FAILURE RELAY.

TER   STATOR TEMPERATURE RELAY.

FFR   FIELD FAILURE RELAY.

DPR   DIRECTIONAL POWER RELAY   E   INSTANTANEOUS EARTH FAULT RELAY

TMR   TIME LAG RELAY   3 PH / S.SW   3 PHASE SELECTOR SWITCH

CCR   CIRCUIT CURRENT RELAY

TR / HR   TRIPPING RELAY HAND RESET

OVR   OVERVOLTAGE RELAY

VV / TM   UNDERVOLTAGE TIMING RELAY

THPU / RE   3 POLE THERMAL OVERLOAD AND PHASE UNBALANCE 1 POLE INSTANTANEOUS EARTH FAULT.

FG / Q   FLAG RELAY (QUALITROL)

Q   QUALITROL RELAY

REF   RESTRICTED EARTH FAULT RELAY

I   INTERLOCK

(KW/KVAr)   KILOWATT KILOVARS METER

(A)   AMMETER

(V)   VOLTMETER

(HZ)   FREQUENCY METER

(COSØ)   POWER FACTOR METER

(SYN)   SYNCHRONISING FACILITY

(KWH)   KILOWATT HOUR METER

(MW)   MEGAWATT INDICATOR   ⟨MW⟩ MEGAWATT CHART RECORDER

(MDA)   DEMAND   MAXIMUM DEMAND AMMETER WITH ALARM CONTACT

(MVAr)   MEGA VAR INDICATOR

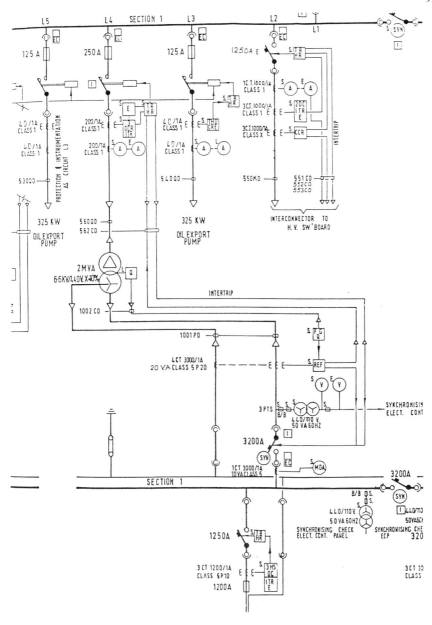

Figure 7.1 Part of fully developed single-line diagram: (a) transformer feeder;
(b) generator incomer (overleaf)

92

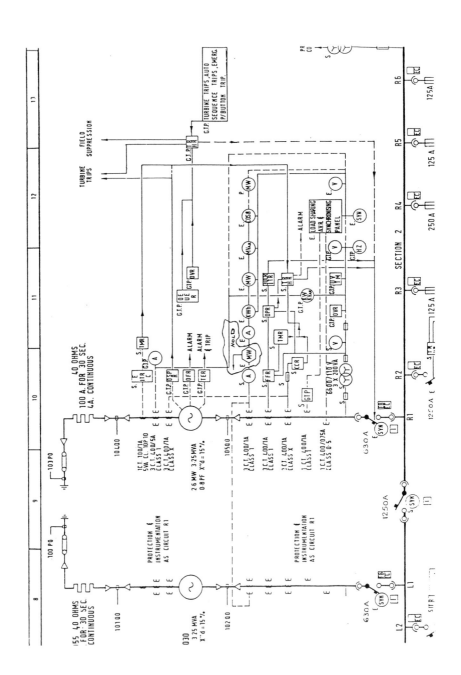

7.1.3.2 Equipment location diagram

In the congested areas of an offshore installation, it is helpful to provide a separate drawing showing equipment locations only. Part of a typical equipment location drawing is shown in Figure 7.2.

7.1.3.3 Cable rack routing diagram

Particularly in the more complex process and power generation modules, cable racking becomes a three-dimensional puzzle. This puzzle needs to be solved in conjunction with the routing of pipework and ventilation ducting in areas already congested by the process or power equipment itself, the structural steelwork of the installation and all the other ancillary equipment such as lighting, communications and instrumentation.

It is no surprise that computer aided design (CAD) systems are used extensively for the draughting representation of such areas, so that clashes between pipework and cable rack routes can hopefully be avoided. Typical illustrations of cable rack diagrams are shown in Figure 7.3. These drawings must also identify the width and number of tiers of rack and tray required between any two adjacent nodes in the cable network. Calculations must be undertaken to size this racking based on the following:

1.   The degree of segregation required (see the section on cable installation to follow).
2.   The number and size of cables passing through this section.
3.   The allowance to be made for growth in numbers of cables during the operational life of the tray or rack. This will depend on the location of the particular section of tray or rack, the age of the installation and the stage reached in the project, but allowances in the region of 300% have been known.
4.   The dimensions of the space available for the racking. If multitier racking is necessary, it is important to ensure that there is sufficient space between each tier for reasonable access. Unless it can be guaranteed that all the cables are of small diameter, a minimum of 300 mm is recommended.
5.   The maximum rack loading in kilograms per metre length of rack as quoted in the manufacturer's catalogue.

The calculation for each section is best done on a prepared calculation sheet similar to that shown in Figure 7.4.

7.1.3.4 Cable routing diagram

For the smaller, less complex systems, racking and cable diagrams may be combined. However, in the more complex arrangements, it is more informative to present the information on two or more separate drawings so that the service, route, identification number etc. of each cable can be easily identified.

Separate diagrams may be produced for each service, such as power, instrumentation, communications, fire and gas etc. The disadvantage with such separation is that there will be less likelihood of spotting a clash if one exists, unless an integrated CAD system is being utilized which can identify such clashes. A typical example of this drawing is shown in Figure 7.5. Cable Transit Sections (Figure 7.5(b)) are required in order to determine the size and number of cable transits.

94

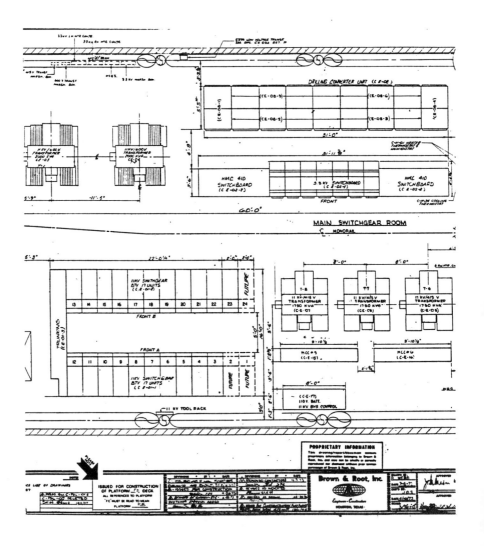

Figure 7.2 Typical equipment location diagram. (Courtesy of BP Exploration Ltd.)

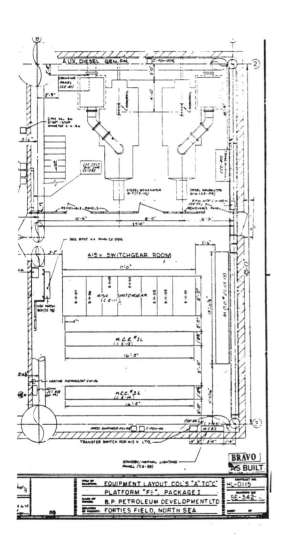

96

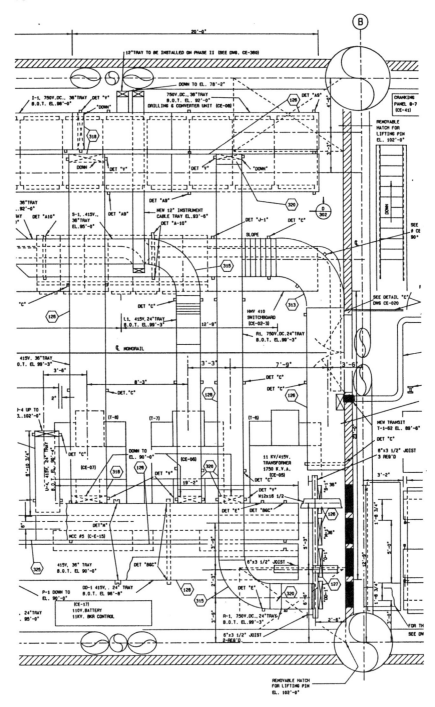

Figure 7.3 Part of a typical cable racking diagram. (Courtesy of BP Exploration Ltd.)

Cable number	Size MM²	Cores	O.D MM	Remarks
2/1	300	1	40.1	
2/2		1		
2/3		1		
2/4		1		
2/5		1		
2/6		1		
2/7	2.5	5	18	
87	4.0	3	18	
88	2.5	5	18	
96	4.0	3	18	
97	35.0	3	34	
99	2.5	5	18	
100A				
187	2.5	37	37	
189	2.5	37	37	
				Total 478.7

Cable number	Size MM²	Cores	O.D MM	Remarks

Proposal/ contract no:	Calculation sheet	No: 0 1 2
Client:		Date 17/12/86
Location		By
Subject: Electrical	Cable ladder infill estimate	Check
Sheet: of		Appr.

POINT ~ B

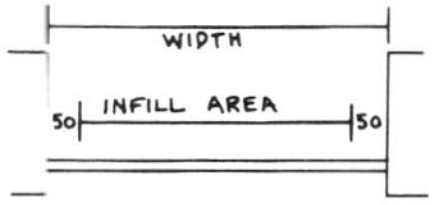

WIDTH
INFILL AREA 50
50

Basis of Calculation

1) Cable schedules submitted for class 2 estimate
2) Infill area having 2 layers of cable
3) Ladder sizes from Swifts catalogue for heavy duty type

Calculation

Additive total of cable diameters = 478
(from loading sheets)

For 50% spare capacity
Total infill area required = 1.5 × 478
= 718

Standard width	Infill area available	Infill area required	% spare capacity	Remarks
150	100	718		
300	400	✓		
450	700	✓	46	
600	1000		109	
750	1300			
900	1600			

Notes

Figure 7.4 Typical cable rack loading calculation sheets

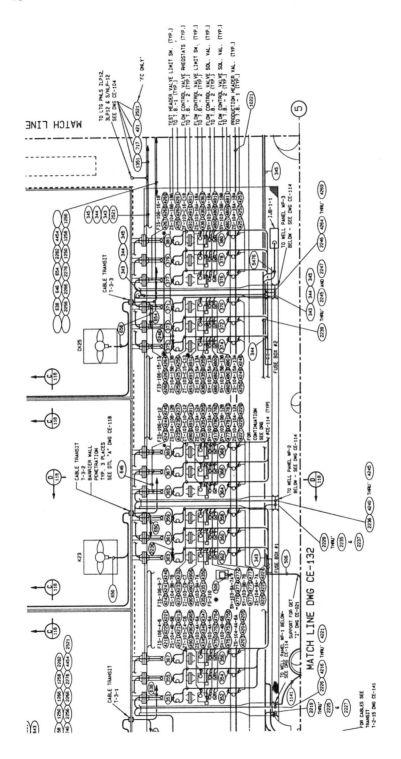

(a)

99

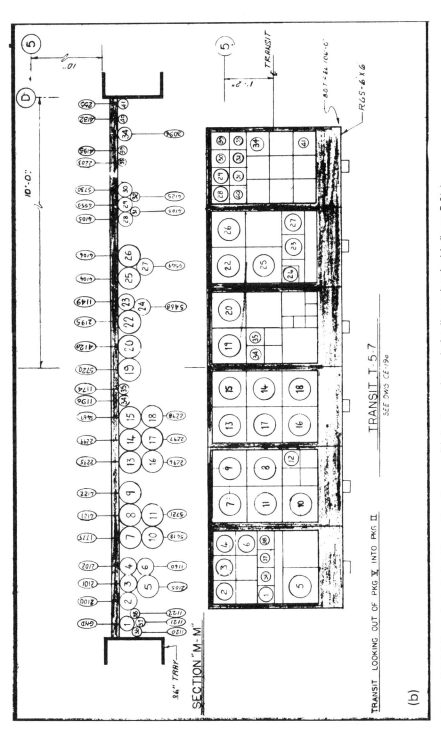

Figure 7.5 (a) Part of a typical cable routing diagram, (b) Section and transit detail associated with figure 7.5(a)

(b)

7.1.3.5 Equipment schematic and block diagrams

These diagrams would normally be produced by the equipment manufacturer, but in some cases the manufacturer's drawings only describe the skid mounted, factory produced unit. These would need to be supplemented by a block diagram showing an overview of the whole platform-wide system and also, if necessary, a comprehensive schematic. Figure 7.6 shows a typical example of an equipment schematic.

7.1.3.6 Panel general arrangement

Before panel wiring and schematic diagrams are produced, either by the manufacturer or by the design contractor, it is necessary to produce a drawing giving arrangement details for the panel. The layout of any control panel is important from both ergonomic and safety points of view, and producing this drawing also assists the designer in the selection of switches, push buttons, lamps, annunciators etc.

7.1.3.7 Panel wiring diagram

Again, these are normally produced by the equipment manufacturer, but in certain instances may require supplementing. Modifications may be required long after the manufacturer's original involvement, and it may not be possible to call upon his assistance again.

7.1.3.8 Termination and interconnection diagrams

When schematic and wiring diagrams have been completed, termination diagrams need to be produced (see Figure 7.7 for an example). These diagrams are the key to platform cable routing because they not only highlight all equipment terminations but also identify all cables and their cores. Once the majority of these have been identified, cable and rack drawings can be produced in detail.

7.1.3.9 Circuit breaker and starter schematic/wiring diagrams

These are similar to other schematic and wiring diagrams, although the design contractor will normally have a greater input to these because they will interface with design information from a number of equipment manufacturers. One manufacturer may be responsible for the main switchboard and generators, in which case there will be less for the design contractor to do. However, this still leaves the design of interfaces between the main switchboard and the controls for all of the large process drives, such as main oil line pumps and gas compressors. These diagrams must include not only the interfaces for the drive controls and instrumentation, but also the interfaces for platform monitoring systems such as ESD and fire and gas systems. An example is shown in Figure 7.8.

7.1.3.10 Junction box and distribution board interconnection diagram

Every offshore installation has a multitude of junction boxes, terminal boxes and distribution boards, and the interconnection of these must be detailed adequately for both installation use and maintenance records. An example is shown in Figure 7.7.

7.1.3.11 Hookup diagram

With offshore labour costs approximately five times higher than those onshore, any means of presenting design information which will assist the

101

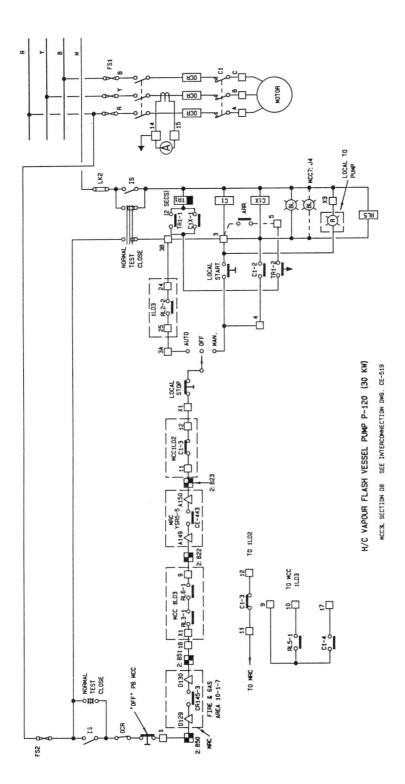

Figure 7.6 Part of a typical equipment schematic. (Courtesy of BP Exploration Ltd.)

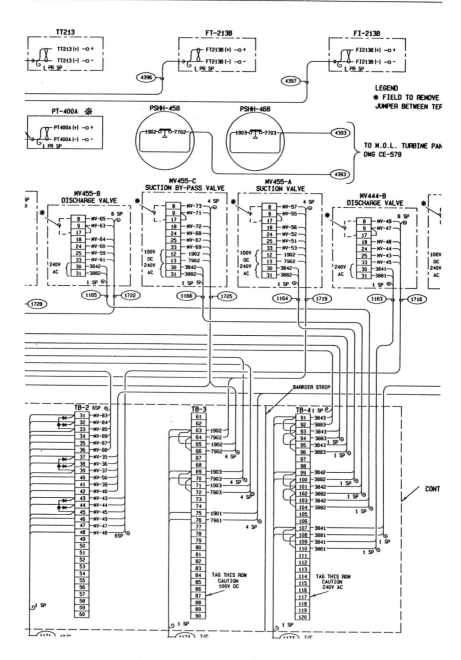

Figure 7.7 Part of a termination and interconnection diagram. (Courtesy of BP Exploration Ltd.)

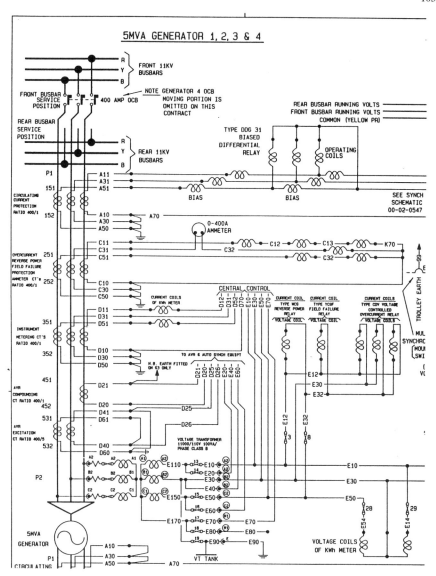

Figure 7.8 Circuit breaker and starter schematic/writing diagram

offshore electrical personnel in their task is worth considering. The hookup diagram is the electrician's equivalent of a motorist's route planning map. It follows an individual circuit for a pressure switch, for example, through all the cabling, back to the control panel and the source of electrical power, identifying every terminal and cable core on the way, just as the motorist's map would concentrate on a description of the route between the departure and destination points.

These drawings tend to be used mainly for instrumentation, and each instrument connection would be drawn on a separate sheet. An example is shown in Figure 7.9.

7.1.4 Databases and schedules

7.1.4.1 Cable schedules
A large offshore installation may have hundreds of kilometres of cables installed, and it is important that, during the design stage, as much information as possible on the identity, route size and type of each cable is retained. This information will obviously be needed for installation, but is also vital should it be necessary to trace, reroute or replace a cable some time later, owing, for example, to fire or mechanical damage. If a particular motor is uprated, it will be necessary to look at the schedule to check whether the existing cable has the capacity for the increased load.

Small project cable schedules may be produced manually using draughting blanks. However, a more practical medium is the computer database, especially during a large project when large numbers of cables are continually being added. For ease of identification, cables should be grouped into service and area. Insertion of extra cables into a schedule would be extremely laborious on a manual schedule but is relatively easy on a database. Other advantages are that duplicate cable numbers cannot exist and that most database programs have powerful sorting and search facilities. An example is given in Figure 7.10.

The cable schedule should also contain a list of cable types, giving a full description of each type, a type code that can be referred to in the main schedule, and a total length for each type. Drum and cutting schedules should also be produced in order to minimize cable wastage during installation. The cable schedule must state how the individual cable lengths were derived. For example, do they already contain a margin for drawing inaccuracy, cutting wastage, installation errors etc.?

7.1.4.2 Electrical equipment schedule
As an aid to draughting and ensuring that every node in the electrical distribution is catered for in the design, an electrical equipment schedule should be produced. This may be combined with a master equipment list later on in the project. Every item should be allocated a client tag number, and the information presented should include the service, location, manufacturer, type and environmental details such as ingress protection (IP) number and suitability for hazardous areas. A motor schedule will give details of the controls and instrumentation required at the motor, at the starter and at any other point of control. It should also indicate the motor full load current, the fuse size and any special requirements such as earth fault relays or thermistors. The schedule may be split into types of equipment such as motors, junction boxes and luminaires. In some cases it will only be beneficial to produce schedules for particular types of equipment, depending on their populations and whether the same information has been produced on another document.

An example of this type of schedule is shown in Figure 7.11. Again, a computer database is recommended for the production of these schedules.

105

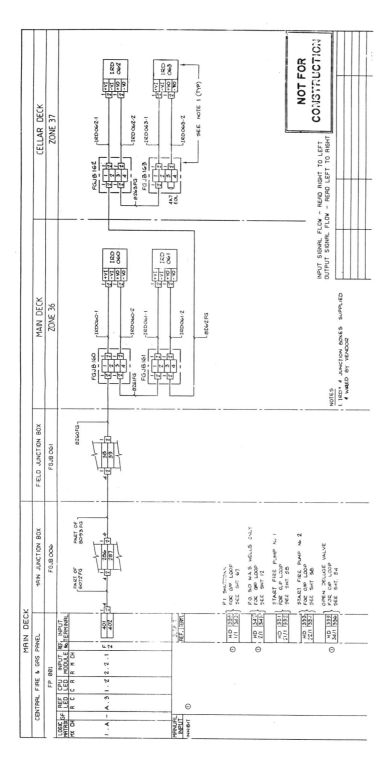

Figure 7.9 Typical hookup diagram

CABLE No.	ROUTE FROM EQUIPMENT		TO EQUIPMENT		CABLE TYPE	No. OF CORES	SIZE mm²	WORKING VOLTS
1	TURBINE GEN. 61	5 MVA	11KV SWITCHGEAR	CB-9	A.1.0	SINGLE	95	11 KV
2	62	5 MVA		CB-10				
3	63	5 MVA		CB-15				
4	11 KV SWITCHGEAR	CB-8	11KV/600V TRANSF. T1	2 MVA	A.1.1	1 3/C		
5		CB-18	T2	2 MVA				
6		CB-6	11KV/3.3 KV T3	3.5 MVA				
7		CB-7	T4	3.5 MVA				
8		CB-19	T5	3.5 MVA				
9		CB-5	11KV/415V T6	1750 KVA				
10		CB-20	T7	1750 KVA				
11		CB-21	T8	1750 KVA				
12	11 KV/600V TRANSF. T1		DRILL CONV. INCOMING CUB INC.#1		A.3.4	SINGLE/C 12	300	600
13	AUX. TRANS.		#1		A.3.3	1-4/C	25	415
14	11 KV/600V TRANSF. T2		#2		A.3.4	SINGLE/C 12	300	600
15	AUX. TRANS.		#2		A.3.3	1-4/C	25	415
16	11 KV/3.3V TRANSF. T3		3.3 KV SWITCHGEAR	CB-32	A.2.1	2-3/C	185	3300
17	T4			CB-34				
18	T5			CB-36				
19	11KV/415V TRANSF. T6		415V	CB-41	A.3.4	SINGLE/C 17	300	415
20	T7			CB-45				
21	T8			CB-47				
22	DEISEL GEN. 66			CB-43		SINGLE/C 11	150	
23	67			CB-49				
24	415 VSWGR	CB-48 CUB. E	M.C.C. NO. 4 IS 53 TIER 'E' UPPER			SINGLE/C 13		
25		CB-44 CUB. C	M.C.C. NO. 4 IS 51 TIER 'E' LOWER					
26	M.C.C. - 1LK1	CB-56	M.C.C. No. 5 COMPT. E1					
27	M.C.C. - 2LA1	CB-57	M.C.C. No. 6 COMPT. A1					
28	TURBINE GENERATOR 61		NEUT EARTHING RESISTOR BOX NER 1		A.1.0	SINGLE	95	11 KV
29	62		2					
30	62		3					
31	NEUTRAL EARTHING RESISTOR #1		EARTHING SYSTEM		A.3.4	3-1/C		
32	NEUTRAL EARTHING RESISTOR #2		EARTHING SYSTEM			3-1/C		
33	NEUTRAL EARTHING RESISTOR #3		EARTHING SYSTEM			3-1/C		
34	11KV/600V TRANSF. 'T1'		NER 'T1'		A.3.4	1-1/C	300	
35	11KV/600V TRANSF. 'T2'		NER 'T2'		A.3.4	1-1/C	300	
36	NER 'T1'		EARTHING SYSTEM		D.I.F.	2-1/C	95	
37	NER 'T2'		EARTHING SYSTEM		D.I.F	2-1/C	95	
38								
39								

Figure 7.10 Part of a typical sheet from a cable schedule. (Courtesy of BP Exploration Ltd.)

GRADE	ROUTE FEET	DUTY	EQUIP. No.	MANKE GLAND CABLE	CABLE ROUTED VIA TRAY	REMARKS
11 KV	375	POWER	C-J-01	D	B7 B5 B2 B3 & B1	RUN CABLES IN 1 TREFOIL GROUP
	350	.	C-J-02	D	B7 B5 B2 B3 & B1	
	280		C-J-03	D	A7 A5 A6 A3 & A1	
	70		T1	F	LOCAL PKG I	
	50		T2			
	85		T3			
	65		T4			
	65		T5			
	75		T6			
	55		T7			
	40		T8			
1000	50			C2		RUN CABLES IN 4 TREFOIL GROUPS @ 3" BETWEEN GROUPS
	50			C		
	65			C2		RUN CABLES IN 4 TREFOIL GROUPS @ 3" BETWEEN GROUPS
	65			C		
3.3 KV	70			F		3" BETWEEN GROUPS
	100			F		.
	95			F		
1000	65			C2		5 TREFOIL GROUPS @ 3" BETWEEN GROUPS (2 CABLES NEUT CONN)
	60			C2		
	60			C2		
	40		C-J-06	C		3 TREFOIL GROUPS @ 3" BETWEEN GROUPS (2 CABLES NEUT CONN)
	40		C-J-07			
	300				A7 A5 A6 A3 & A1	2 TREFOIL GROUPS @ 3" BETWEEN GROUPS (2 CABLES NEUT CONN)
	325				B7 B5 B2 B3 & B1	
	60				LOCAL PKG I	
	65					
11 KV	18	EARTHING		D	B7-1 PKG VII	NEUTRAL CONNECTION
	18			D	B7-2	
	18			D	A7-1	
1000	16			C	LOCAL	EARTHING
	11			C		
	16			C		
	40			C2	LOCAL PKG I	
	40			C2		
	50			C		
	50			C		

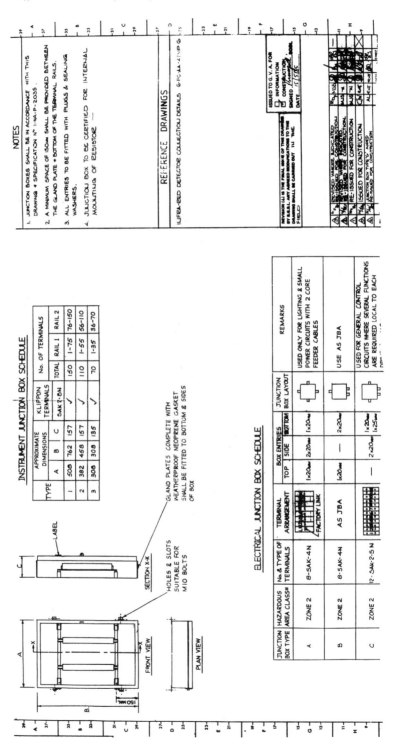

Figure 7.11 Typical sheet from an electrical equipment schedule for junction boxes. (Courtesy of BP Exploration Ltd.)

7.1.4.3 *Plug and socket schedule*

These schedules are often required for diving and subsea equipment, as such equipment tends to be a prolific user of special underwater connectors. Each connector may have over 100 connections. Pin current ratings may not all be identical on the same plug, and it may also be necessary to parallel several pins in one circuit in order to obtain the required rating. It may be necessary to monitor circuits for continuity and also for earth leakage. To further complicate the matter, every socket of identical size will require a different orientation to prevent plugs being mated with the wrong sockets.

Some form of schedule is essential in order to keep track of circuit routeing through the various connectors. Each connector should be provided with a separate page (or pages) in the schedule, and this will need to identify the service, the plug and socket manufacturer and catalogue number, and the orientation between plug and socket. Each circuit should then be listed, with details of cable cores terminated in the plug and socket, circuit rating, purpose etc.

7.1.4.4 *Electrical load list*

As with any project where electrical power will be required, a continuous review of the project loading is necessary. This is particularly important where the electrical system is isolated and dependent on its own power sources to support the load. The preparation of an accurate load list often becomes a priority early in the project, since the less accurate the system load figures, the more risk involved in purchasing suitable generators. In any case a good margin should be kept over estimated loadings if generator outputs are not already restricted by weight and space limitations on the installation. The schedule should indicate connected and diversified loads for all the system and generation operational states envisaged. A typical load list is shown in Figure 7.12.

7.2   Material and equipment handling and storage

Adequate secure, warm, dry storage needs to be provided for equipment prior to being called for offshore. The storage facility should be local to the offshore supply base if construction delays are to be avoided.

A goods inspection procedure is important, as it should detect damaged or incorrect items which can then be repaired or replaced before they are sent offshore. An item of equipment whose damage or unsuitability remains undetected until it is being installed can play havoc with the construction programme.

The contractors carrying out the installation work will need to store materials offshore, so suitable containers will need to be located on the installation. The areas where such containers are stored are known as laydown areas. Laydown areas need to be adjacent to platform cranes in order that containers and equipment can be easily loaded off and on supply vessels. Crane operations need to be studied to ensure that the risk of

110

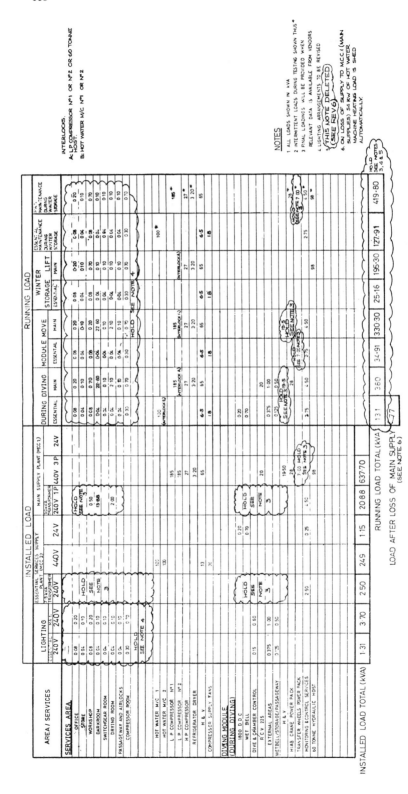

Figure 7.12 Part of a typical sheet from an electrical load list

dropping or impacting equipment or containers on modules, particularly those containing hazardous production equipment, is minimized.

Every item sent offshore should be clearly identified and marked with its gross weight. This reduces the incidence of loads being dropped by overloaded cranes, or of mysterious packing cases arriving on some other oil company's offshore installation when the equipment inside is vital to the next stage of your client's construction programme. Once equipment has reached the offshore installation, particularly if heavy structural work is going on, it must be protected from physical damage, shock, vibration, ingress of dust, moisture, welding sparks and any other foreign matter until it is permanently protected.

The manufacturer's storage and preservation procedures must be adhered to if warranties are to remain valid.

7.3 Erection procedure

An installation specification should have been produced in the design office, and copies of this should be made available to those carrying out the installation and to installation inspectors. As well as an installation general code of practice, this document should also contain a series of blank test sheets for recording all electrical equipment precommissioning tests.

The arrangements for preservation of the equipment will need to continue while erection is completed. It is important to produce or obtain from the manufacturer a written erection procedure well before the contractor starts work. This will allow the procedure to be checked for compatibility with the relevant construction package and for any operation that might be dangerous to personnel or risk damaging the equipment. The procedure should identify every item of plant that the installation contractor will need to install the equipment, so that space may be allocated for it and, if it is a source of ignition, electromagnetic interference etc., the relevant precautions can be taken in advance.

When cables are being installed or electrical equipment commissioned, the installation contractor should be required to mark up a set of drawing prints showing any changes to cable routes, terminations etc. which prove necessary. These 'as-built' drawings are then returned to the design office for review and, if they are accepted, the changes should be incorporated in the drawing masters. Providing the work is done to the inspectors' satisfaction, the contractor will receive a certificate of acceptance, which he will need for invoicing purposes.

The following are practical notes on the installation of particular equipment. Commissioning and precommissioning tests are covered in Chapter 13.

7.4 Switchgear and motor control centres

Although the majority of the installation work is usually done by the offshore contractor, a technical representative from the switchgear manufacturer should be present to ensure that the work is to the manufacturer's satisfaction and that any problems can be rectified quickly.

The installers should be completely familiar with the equipment and all the information regarding its installation. They should certainly not see it for the first time as it is removed from the packing case at the installation site. This is particularly important with regard to foundation and supporting steelwork arrangements. These must be in place, and of the correct dimensions, before installation of the equipment itself can be attempted.

Once the framework is in place, correctly aligned, levelled and rigidly fixed, switchgear and control gear cubicles can be bolted together in the required order as shown on the design drawings provided. Identification labels should be checked to ensure that they correspond to the equipment and the associated drawings.

All the copper connections for busbars and risers should then be made using torque limiting tools, under the supervision of the manufacturer's representative. Where the manufacturer recommends or provides special tools, only these should be used. Ductor resistance testing of busbars can then be carried out. Any bus ducting between the switchgear and associated transformers or generators should be installed at this time, if this is possible. This should avoid any stressing of connections. If the switchgear is being installed in a module at a fabricator's yard prior to being shipped offshore in the module, then any transit packing removed should be retained, as this will be needed for repacking in order to prevent any shock or movement damage during the shipment offshore. All transit packing should be removed, and the equipment properly cleaned prior to testing.

A check for correct mechanical operation will then need to be made. This should prove that:

(a)   withdrawable trucks, cubicles and interlocking devices function correctly;
(b)   circuit breaker mechanisms, isolators, switches and relays are free to open and close properly.

By this time, the tripping battery and charger should have been installed and commissioned. Nevertheless, it is worth checking that the tripping batteries are fully charged before going any further.

When all the circuit breakers have been opened and closed electrically, a start can be made on primary injection testing of relays, and megger and high-voltage insulation checks can be completed, since these are more easily carried out before any cabling work commences.

Before any incoming or supply cables are connected, all the required earthing copperwork and cabling should be completed and tested.

The switchboard should now be ready for connection to incoming supplies. Cabling checks are discussed in a later section, and commissioning in Chapter 13. When the switchboard has been energized, the contactors may be checked for electrical operation. Supply circuits are then commissioned on a piecemeal basis as part of the commissioning procedure for the item of equipment being supplied.

7.5   Distribution transformers

Distribution transformers are heavy devices, and it may have been necessary to strengthen a particular section of module floor or platform deck at the installation site. The installer should ensure that all transformers are placed on a flat, level, previously prepared base, in the location shown on the relevant design drawing.

Once a transformer is installed, it may be necessary to fill it if it is a fluid filled type. Some types of insulation fluid are toxic and therefore special precautions will need to be taken.

The manufacturer's representative should be called in after all the transformers have been located and filled, to assist and witness the final checking and testing. Insulation tests should then be completed. After cabling, a second insulation test of all windings and connections should be carried out.

On installations having two or more transformers connected to a single switchboard, a phase polarity check should be carried out across the associated bus section.

7.6   Motors and generators

In most cases, rotating machinery will be installed as a complete skid mounted package. Whether they are installed as separate items or not, great care must be taken to ensure that the motor and driven machinery are correctly located and aligned after installation. The equipment must be slung or lifted from prescribed lifting positions such as eyebolts where these are available. A normal rule of thumb is that the lifting orientation of the machinery should be identical to its operating orientation, i.e. normally lift in the upright position.

During the entire installation procedure, it is important that any tools or materials supplied specifically for the installation of the particular machine should be used. It is not uncommon for such items to be left in the equipment packing case or otherwise overlooked, and for installation to then go ahead using improvised methods which may jeopardize the success of the project and possibly the safety of offshore personnel.

Protective films on machined surfaces should be cleaned off with the recommended solvent. This cleaning should take place immediately prior to installation, where plinths are in contact with baseplates etc., and prior to commissioning for such items as motor shafts. Shafts and bearings need to be kept clean and covered whenever work is not proceeding.

Shims supplied with the equipment should be used wherever possible for levelling. Shims should be of similar area and shape as the machined surfaces of the equipment footings for which they are used, and the maximum thickness of shim should always be used so that the minimum number of shims are positioned under each footing. Where separate bearing pedestals are supplied, these should be positioned, levelled and lined up prior to the installation of other machinery components. White metal bearings should be checked to ensure adequate bedding using engineer's blue or some other suitable method of indication.

Coupling faces should be truly parallel and level. This will need to be checked with each shaft rotated to several different positions. Coupling bolts or flexible connections should be properly fitted without damage and all nuts tightened and locked before running the machinery.

When larger machines, and particularly synchronous machines, are installed, the manufacturer's instructions concerning insulation must be carefully followed to prevent circulating currents flowing in the machine frame. Where the machine frame is installed in sections, conductive bonding between each section will need to be established, if necessary by separate bonding conductors. This is particularly important when the motor is sited in a hazardous area, since otherwise sparking may occur between sections.

The larger, heavier machines should be provided with permanent runway beams and horizontal and vertical screw jacks to facilitate alignment. Alternatively, suitable jacking points can be provided for temporary jacks. The use of block and tackle arrangements or tirfors is time consuming and likely to increase the risk of injury to installers.

Before commissioning, machines should be cleaned; if they are of the open type, dust and dirt should be blown out. Commutators and slip rings will need to be checked for deterioration during transit and storage, and cleaned if necessary. Carbon brushes will need to be checked for freedom of movement, and springs checked for correct pressure. If machines have been stored for a long period, the bearing grease should be inspected; if deterioration has occurred, the bearings should be thoroughly washed, dried and regreased using methods recommended by the particular bearing manufacturer. Where oil lubricating systems are employed, bearing oil rings, flow switches, pressure switches, pumps etc. must be checked to ensure they are fully operative before any running of the machine is attempted.

Machine shafts will require to be rotated by hand or barring gear to ascertain that no foreign body is either inside the machine or between the external fan and its protecting cowl. Checks should also be made to ensure that ventilating air ducts are clean and clear of obstruction. Where safety guards are provided with the machinery, these should be fitted prior to any rotation of the equipment under power. The guards should be constructed so as to facilitate removal. The location of both the equipment and the guards on the equipment should be such that access for inspection and maintenance is practicable and safe. Both guards and couplings between motors and machinery should be removable without requiring removal of the motor or the driven equipment. Where guards are over belts and pulleys, a check should be made that the belt does not slap the guard at any speed. This is particularly important with variable speed drives.

Before any rotating equipment is commissioned offshore, the following precautions should be taken:

1.   All necessary tools and ancillary equipment should be conveniently to hand but not so placed as to pose a safety hazard.

2. Notices relating to fire fighting procedures and treatment of electric shock and burns should be displayed in switchrooms and control rooms.
3. All normal working and warning notices for the equipment should be clearly visible.

7.7   Lighting and small power

During the design phase, calculations should have been carried out in order to fix types, light outputs and locations of all the luminaires to be installed. If the required lighting levels are to be obtained, the installers must wherever possible adhere to the design details provided. This will not always be possible, particularly in congested areas, and to avoid shadow or obstruction of walkways etc. an optimum location should be chosen by the installer.

For obvious reasons, luminaires are conspicuous; their acceptable appearance should be preserved by careful installation, especially in accommodation areas. Rows of fittings should be installed accurately in a straight line, and fastenings and suspensions rigidly set up so as to avoid distortion by handling during normal maintenance. The colour rendering of tubes and lamps should be consistent, and suitable for the area where they are installed.

It is important both for balanced loading of phases and for ease of future .dentification that luminaires, power sockets and distribution boards are wired in accordance with the design circuit schedules provided. Care should be taken to ensure that polarities are correct when making connections to switches, convenience sockets, lamp holders and similar items. When cabling up to swivelling floodlights, enough cable should be provided to allow the floodlight to be swivelled a full 360 degrees.

In some cases, luminaires with integral emergency batteries are provided with batteries unfitted. The batteries should be fitted as soon as possible to avoid deterioration, especially if the battery housing is required to seal the luminaire enclosure. Immediately after installation, the luminaires should be provided with a suitable electrical supply (even if this supply is temporary) for a continuous period of 80 to 100 hours to ensure their batteries are fully charged.

7.8   Secure power supply systems

7.8.1 Batteries

Under normal circumstances, batteries will be shipped filled and discharged. Sealed batteries will be shipped filled and charged. If batteries are shipped unfilled, appropriate safe filling facilities will need to be made available offshore.

Battery racking, cell units, connecting conductor links and associated nut, bolt and washer kits should be inspected for correct type and quantity on arrival. Any damaged components or shortages will need to be made up before installation can be completed. Unsealed batteries will require either filling or topping up depending on the condition in which they were supplied.

It is important that batteries are assembled as shown in the drawings supplied. Incorrect configuration may provide the wrong output voltage and/or discharge duration and, in any case, links will have been provided in the correct numbers and sizes for a particular configuration. Link and cable lug bolts should be tightened using a torque limiting tool, particularly with sealed cells whose terminal posts snap off at very low torques. On unsealed batteries particularly, all terminals and connections should be liberally greased with petroleum jelly or similar substance according to manufacturer's recommendations.

7.8.2 Battery chargers and inverters

Cubicles should be positioned to ensure that a free flow of cooling air is available and that the ventilation entries and exits are clear of obstructions. An all-round clearance space of at least 150 mm is recommended.

Before commissioning, all cabling connections, circuit breakers and fuses should be checked for correct rating and operation. The manufacturer's representative should be present to inspect the equipment to his satisfaction before supervising its commissioning.

7.9   Communications

7.9.1 Public address systems

The public address (PA) system is vital for communication of hazards and the calling of staff in the field, so it is often necessary to commission a part of the system as soon as it is installed and to minimize disruptions to its operation whilst installation of the rest of the system continues. This is quite possible provided a powered amplifier rack is available at an early stage in the proceedings.

For reliability, it is a requirement that in every area the PA circuits are duplicated. Therefore at least two circuits of loudspeakers will be required on any installation; in practice more will be required, since it will be necessary to mute speakers in sleeping areas during normal operating conditions. On large installations, more than two amplifier circuits will be required for reasons of loading and as a maintenance facility. As the system will be required to operate during abnormal conditions such as serious gas leaks, the junction boxes, changeover switches, loudspeakers and isolators should all be certified for use in zone 1 hazards, even when sited in areas classified as non-hazardous during normal conditions.

7.9.2 Telephones

The installation of offshore PABX systems is similar to that for onshore systems but with the addition of the following.

First, in hazardous areas the equipment, including the telephone itself, must be certified for the zone of hazard concerned. Telephone instruments are normally of the flameproof type in such conditions.

Secondly, in noisy areas such as machinery rooms, the telephone instrument will initiate, via a relay box, horn and light signals, to indicate when the instrument is being called. The relay box, horn and signal lamp should be mounted near to the associated telephone instrument. In some cases, a relay box may control more than one horn and lamp arrangement depending on the size and degree of congestion within the module concerned.

7.10 Cable support systems

In the design phase, a great deal of care should have been taken to ensure that the correct sizes, configuration and routes have been shown on the cable and racking arrangement drawings. If route clashes with pipework or overloading of supports are to be avoided, it is vital to install both support systems and cable strictly according to design drawings. However, there is always the possibility that a design error has been made. Thus the installer should carefully check the locations as shown on the drawings to determine if any conflict exists between the new cable route and any other equipment, steelwork, piping, ducting etc., and whether, when installed, the way will cause a hazard to personnel, obstruct accessways or prevent the installation or removal of equipment.

On minor routes, not detailed on design drawings, where it is necessary to run cable support systems, it is important that separate racks or trays are used for the following categories in order to prevent electromagnetic interference between cables:

(a) medium-voltage AC (above 1 kV) with associated control cables;
(b) low-voltage AC (240 V to 1000 V) with associated control cables; DC power cables, associated control cables and 110 V DC/254 V AC instrument cables;
(c) instrument, signal and alarm cables; telephone, communications, and fire and gas cables.

Where practicable a minimum segregation distance of 1 metre should be maintained. Crossovers should be kept to a minimum and as near perpendicular as possible.

Some vital circuits associated with fire and gas, ESD or other safety related systems may have supply and/or signal cables duplicated. Such cables should be run on separate routes so that a single fire cannot destroy both cables.

7.10.1 Support steelwork

Drawings showing details of standard steel supports and brackets should be supplied as part of the design drawing package. If not, perhaps because a particular situation was not envisaged, then the installer will need to sketch a suitable arrangement and obtain approval for its use. The work will be made easier if one of the proprietary steel framing systems such as Unistrut or Leprack is used.

The steelwork is usually stainless or hot dip galvanized mild steel, and should be free of sharp edges and burrs likely to damage cables. Nuts, bolts etc. may be stainless or cadmium plated mild steel, and ISO metric threads should be used throughout. Once installed, the whole arrangement may be given the standard paint finish before the cables are positioned.

Supports for horizontal tray or rack should be spaced according to the type, width and estimated maximum loading, but should never exceed 3 metres. Supports for vertical tray or rack should be spaced at approximately 1 metre intervals, and should provide a clear space between the rack and the structure of at least 400 mm to allow for pipe lagging.

Steel or concrete members forming part of the module or installation structure must not be drilled or welded to provide a fixing point for supports, unless written permission has been obtained. Such drilling or welding could weaken the structural integrity or reduce the seaworthiness of the installation.

7.10.2 Cable tray

A variety of different types of cable tray are in use offshore, and they are made from a number of different materials. The following are two common types found offshore:

1. Heavy duty Admiralty pattern. This may be stainless steel or high-quality Corten A which has been hot dip galvanized several times.
2. Heavy duty reverse flange. This is much stronger mechanically than the equivalent Admiralty pattern because of the doubling over at the edges (reverse flange). Again, the material may be stainless steel or hot dip galvanized Corten A.

Epoxy coated mild steel types should be avoided, as they will deteriorate quickly once any damage to the coating occurs.

7.10.3 Tray installation

The following points should be considered when specifying and installing tray:

1. Manufacturers' bend, tee and crossover sections should be used rather than sections fabricated on site from straight sections. Site fabricated sections tend to be weaker mechanically, and more likely to damage cables by burring or unfinished metal. If it is necessary to cut a section of tray, it should be cut along a line of plain metal rather than through the perforations.

2. Admiralty pattern and reverse flange tray should not be used in the same area, as the two types cannot be easily joined together. Reverse flange tray is normally used for external areas only.
3. At 25 metre maximum intervals along the tray, it should be bonded to the platform structure. With most types of tray, earth bonding continuity will be provided between sections by the tray itself, although some operators require a braided bonding connector to be used.

7.10.4 Ladder rack

There are a variety of rack types used offshore. All are based on a ladder design with various standard components, which allows any three-dimensional configuration, including multitier, to be built up 'Meccano fashion'. If stainless steel rack is to be used, great care will need to be exercised in quality control since there is a tendency for distortion to appear in some production batches.

If glass fibre reinforced plastic (GRP) ladder is used, the stronger 'pulltruded' type is recommended. As there is some fire risk associated with this type of tray, its use should be restricted to lower areas of the installation where sea spray is likely to cause corrosion with metal racking. For the same reason, long vertical runs of GRP rack are not recommended.

7.11 Cables

7.11.1 Selection

Cables used on any offshore installation will require the following attributes:

1. They should have stranded copper conductors for smaller cross-section and better flexibility.
2. They should be voltage and fault rated for the system in which they are operating.
3. They should be rated for normal maximum current flows of the circuit without exceeding the maximum conductor temperature or the temperature class limit if passing through a hazardous area.
4. If they are involved in a fire, smoke and acid fume emission should be low. The insulation compound used should have an oxygen index of at least 30.
5. They should normally be armoured for mechanical protection. Braided or wire armouring may be used; the braided type is slightly more flexible but is often more difficult to gland.

These criteria cannot be rigidly applied, however, since special cables will be needed for instrumentation, diving umbilicals, downhole pumps etc.

The cables designed for the most onerous duty are those for electric fire pump supplies. These fire survival cables are designed to continue to supply current after an hour at a temperature of 1000°C, followed by hosing down with high-pressure water jets whilst being hit with a hammer. The difficulty lies in finding a suitable support system that will survive the same treatment!

The most common cable type in use is the EPR/CSP type similar to that found in merchant shipping, which meets the high oxygen index and fire retardant requirements of BS 6883 and IEC 93-3.

Cables must be sized to allow for circuit current and maximum voltage drop as with any onshore system. AC cables can be sized using a chart generated by spreadsheet, as shown in Figure 7.13. Care will still need to be exercised to ensure that close excess protection is still applied and that cable fault ratings are adequate.

Although the IEE Wiring Regulations (sixteenth edition 1991) do not apply to offshore installations, they should be used for guidance in the sizing of conductors, unless more specific information is made available in a future revision of the IEE Recommendations for the Electrical and Electronic Equipment of Mobile and Fixed Offshore Installations (first edition 1983).

7.11.2 Installation

1. Cables should be routed in such a way as to facilitate maintenance and the installation of additional cables with minimum need for expensive scaffolding. For example, a cable route located under a deck, so that there is no immediate access and the sea is directly below it, is not recommended. The cost of a small cabling modification in such a location would be overshadowed by the enormous scaffolding bill.
2. Cable routes should avoid known fire risks where this is possible. Cables to main and standby machinery should be run on separate routes.
3. Cables should be laid parallel on cable ladders and tray in a neat and orderly fashion.
4. Where heavy three-phase currents are carried, cables are usually single-core. Single-core cables of the same conductor cross-section usually have a higher fault rating than their three-phase equivalents, and are easier to install, having a lower weight per metre and a lower bending radius. To avoid eddy currents being induced in local steelwork, such cables must be run in a trefoil configuration. In some situations the cables must be run in a flat profile because of space limitations, in which case a balanced arrangement must be adopted which still avoids the promotion of eddy currents. Support or other steelwork must not pass between phases, as the steel will heat up owing to hysteresis loss. Where two or more trefoil cable groups run on the same route, they should be at the same horizontal level with a clear space between the groups of at least one cable diameter.
5. Cable bends should not be tighter than the minimum bending radius specified by the manufacturer. Drip loops should be provided at external cable terminations. These will also be useful if the cable has to be reterminated at some later date. In any case, straining of cables at cable glands should be avoided, and cables should be perpendicular

CABLE SELECTION FOR 415 VOLT MOTORS
STARTING CURRENT = 6#FLA

Frequency = 60Hz
CABLE   SWA/EPR/CSP

KW	FLA AMP	CABLE X-SECTION	CABLE AMP	DISTANCE METRES	VOLT DROP PER METRE	DERATE FACTOR	POWER FACTOR	SIN@	RESISTANCE PER METRE	REACTANCE PER METRE	VOLTAGE KILOVOLTS	reactance @ 50Hz
5.5	9.002	1.500	22.00	75.084	1.105	0.409	0.850	0.527	0.0138	0.000164	0.415	0.000136
5.5	8.897	2.500	33.00	136.454	0.608	0.270	0.860	0.510	0.0076	0.000151	0.415	0.000125
7.5	12.133	1.500	22.00	55.079	1.507	0.551	0.860	0.510	0.0138	0.000164	0.415	0.000136
7.5	12.133	2.500	33.00	100.066	0.829	0.368	0.860	0.510	0.0076	0.000151	0.415	0.000125
11.0	17.794	2.500	33.00	68.227	1.217	0.539	0.860	0.510	0.0076	0.000151	0.415	0.000125
11.0	17.794	4.000	44.00	108.986	0.762	0.404	0.860	0.510	0.0047	0.000149	0.415	0.000124
15.0	24.265	2.500	33.00	50.033	1.659	0.735	0.860	0.510	0.0076	0.000151	0.415	0.000125
15.0	24.265	4.000	44.00	79.923	1.039	0.551	0.860	0.510	0.0047	0.000149	0.415	0.000124
18.5	29.927	6.000	58.00	97.145	0.854	0.516	0.860	0.510	0.0031	0.000142	0.415	0.000118
15.0	24.265	10.000	77.00	199.031	0.417	0.315	0.860	0.510	0.0018	0.000140	0.415	0.000116
22.0	35.180	6.000	58.00	81.787	1.015	0.607	0.870	0.493	0.0031	0.000142	0.415	0.000118
22.0	35.180	10.000	77.00	135.967	0.610	0.457	0.870	0.493	0.0018	0.000140	0.415	0.000116
30.0	47.973	10.000	77.00	99.709	0.832	0.623	0.870	0.493	0.0018	0.000140	0.415	0.000116
30.0	47.973	16.000	103.00	155.021	0.535	0.466	0.870	0.493	0.0012	0.000131	0.415	0.000109
37.0	59.166	16.000	103.00	125.693	0.660	0.574	0.870	0.493	0.0012	0.000131	0.415	0.000109
37.0	59.166	25.000	135.00	192.905	0.430	0.438	0.870	0.493	0.0007	0.000124	0.415	0.000103
45.0	71.141	16.000	103.00	103.644	0.801	0.691	0.880	0.475	0.0012	0.000131	0.415	0.000109
45.0	71.141	25.000	135.00	159.272	0.521	0.527	0.880	0.475	0.0007	0.000124	0.415	0.000103
55.0	86.950	25.000	135.00	130.313	0.637	0.644	0.880	0.475	0.0007	0.000124	0.415	0.000103
55.0	86.950	35.000	165.00	175.520	0.473	0.527	0.880	0.475	0.0005	0.000122	0.415	0.000101
75.0	118.569	35.000	165.00	128.715	0.645	0.719	0.880	0.475	0.0005	0.000122	0.415	0.000101
75.0	118.569	50.000	200.00	167.849	0.494	0.593	0.880	0.475	0.0004	0.000120	0.415	0.0001
90.0	142.282	50.000	200.00	139.874	0.593	0.711	0.880	0.475	0.0004	0.000120	0.415	0.0001
90.0	142.282	70.000	250.00	191.506	0.433	0.569	0.880	0.475	0.0003	0.000117	0.415	0.000097
110.0	173.901	70.000	250.00	156.686	0.530	0.696	0.880	0.475	0.0003	0.000117	0.415	0.000097
110.0	173.901	95.000	300.00	198.604	0.418	0.580	0.880	0.475	0.0002	0.000114	0.415	0.000095
132.0	208.681	95.000	300.00	165.503	0.502	0.696	0.880	0.475	0.0002	0.000114	0.415	0.000095
132.0	208.681	120.000	360.00	202.774	0.409	0.580	0.880	0.475	0.0002	0.000112	0.415	0.000093
160.0	252.946	120.000	400.00	192.407	0.431	0.632	0.880	0.475	0.0001	0.000112	0.415	0.000093
160.0	252.946	185.000	460.00	223.580	0.371	0.550	0.880	0.475	0.0001	0.000112	0.415	0.000093
200.0	316.183	185.000	460.00	178.864	0.464	0.687	0.880	0.475	0.0001	0.000112	0.415	0.000093
200.0	316.183	240.000	550.00	211.013	0.393	0.575	0.880	0.475	0.0001	0.000111	0.415	0.000092
250.0	395.229	300.000	630.00	191.553	0.433	0.627	0.880	0.475	0.0001	0.000110	0.415	0.000091

Figure 7.13 Spreadsheet generated cable sizing chart: for basis see Section 9.6.3

121

to gland plates for a minimum of 100 mm before entering the gland. Expansion loops will also be required across the expansion gaps necessary between module walls and support steelwork.

6. Some cable insulation materials tend to become brittle with lower temperatures. This is particularly the case with PVC compounds. Under low-temperature conditions, care needs to be taken when installing most cable types. It is desirable that cables with PVC insulation should not be installed directly at ambient temperatures of 5°C and below. In such conditions, the cable should be stored for at least the previous 24 hours at a temperature of 20°C or more.
7. Cables should be clamped or cleated to rack or tray with suitable ties. The usual arrangement is to use nylon 12 ties (not nylon 66 as it becomes brittle in the offshore environment), supplemented at suitable intervals (depending on the weight) by EVA coated stainless steel banding or cable cleats of the correct diameter. Ties should be non-magnetic for single-core cables cleated in flat formation. Recommended cleating/banding intervals are shown in Table 7.1.
8. Where trefoil cleats are used for single-core cables, stainless steel banding should also be fitted for fault current bracing at 1 metre intervals, with a band always located close to each cleat.
9. The positive and negative cables forming a DC circuit should be run side by side as far as possible to reduce magnetic effects.
10. Spare cores in multicore cables should be coiled back for future use, not cut off at the gland.

Table 7.1 Recommended stainless steel cleating/banding intervals

Overall diameter and type of cable	Maximum spacing of SS banding (mm)	
	Horizontal run (to 60° from horiz.)	Vertical run (to 30° from vert.)
Single-core cables in trefoil formation	2000	1500
Multicore cables (above 35 mm OD)	1500	1000

7.11.3 Transits, glands and connectors

Wherever cables have to penetrate a wall or enclosure which must remain gas tight or has a fire protecting function, it will need to be installed using a transit, gland or connector.

7.11.3.1 Transits
Transits consist of a steel frame which is welded or bolted on to the wall or floor where a suitable slot has been cut. Some transit frames are split so that they can be placed round existing cables without the need for the cables to be reinstalled. Where the cables pass through the frame, they are individually enclosed in shaped flexible packing blocks as shown in Figure 7.14. Each layer of cables and surrounding blocks is anchored into the frame by a stayplate. To fill up any unused space, blank blocks are inserted

Figure 7.14 Illustration of a typical cable transit. (Courtesy Hawke Cable Glands Ltd.)

and the whole assembly is compressed down by a compression plate and threaded bolt. The space left at the top is then filled by an end packing which is fitted with compression bolts to expand sandwiched flexible blocks and complete the sealing process. The resulting arrangement provides an effective and reusable method of penetrating a steel bulkhead or deck, and can withstand gas pressures of up to 4 bar. Good fire tightness properties can be obtained when the transits are used in conjunction with mineral wool or compounds such as Mandolite 25 which cover the entire fire wall.

7.11.3.2 Glands
Glands are normally of the flameproof type with inner and outer seals, as shown in Figure 7.15. Flameproof barrier glands may be used, where the inner seal is replaced by a setting compound which forms a seal round individual cores. The benefit of being able to use more economical cable constructions through the use of barrier glands is usually outweighed by the disadvantage that installation, particularly compound filling, is more difficult, and the cable can only be freed from the gland by cutting it off at the entry point. In either case the gland must be correctly assembled and installed (without cutting or otherwise tampering with its components) if certification is to remain valid.

7.11.3.3 Connectors
Connectors of various types may be used, particularly for diving and subsea applications. Some connectors are designed to be connected with circuits live underwater. Some may be hazardous area certified, although

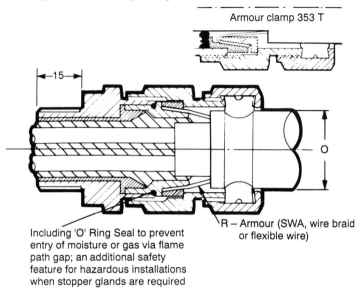

Armour clamp 353 T

←15→

O

Including 'O' Ring Seal to prevent
entry of moisture or gas via flame
path gap; an additional safety
feature for hazardous installations
when stopper glands are required

R – Armour (SWA, wire braid
or flexible wire)

Figure 7.15 Sectional diagram of explosionproof cable gland. (Courtesy Hawke Cable Glands Ltd.)

in this case the circuits would need to be intrinsically safe or provided with some means of preventing live disconnection.

7.12   Bus ducting

The transmission of heavy electrical currents over short distances and in congested areas using cables presents two main difficulties. First, cables of cross-sectional area up to $630 \, mm^2$ are available from manufacturers' standard ranges. However, to cope with loadings of up to several thousand amperes, three or more such large cables per phase may be required. The bunching of such a number of large cables in a congested space requires that their current carrying capacities need to be heavily derated. Secondly, the bending radii of such cables is large, leading to difficulties in installation and termination.

In order to overcome such problems, bus ducting may be used. This has the benefits of high current carrying capacity and the inherent facility for the fabrication of right angle or other bends to order. The bus ducting used must be dust and moisture protecting to at least the same standard as the switchgear to which it is connected, or IP66 standard if it is necessary to run the duct outside the switchgear module. Accurate measurement is required to ensure a close fit and to avoid stressing of the duct or terminating equipment. Unless there is certainty that it is not possible for external ducting to be live during a gas release incident, it is recommended that such ducting is hazardous area certified.

7.13 Control stations, junction boxes and distribution boards

It is important that this type of equipment is suitable for the offshore environment, particularly with regard to the following:

1. In order to be convenient for operation and maintenance, such equipment is often placed in relatively exposed positions, and therefore requires to be both highly resistant to corrosion and mechanically robust. Cast or stainless steels are commonly used; aluminium alloys tend to give corrosion problems. Some cast steel enclosures may also be prone to corrosion, and a deep galvanized finish is recommended. Polycarbonate and GRP enclosures are replacing the metal types in most duties, but should be positioned with care, as they do not often survive a swipe with a scaffolding pole or similar hazard.
2. Isolators, miniature circuit breakers, some push button controls and other switching devices located in the field will require some means of locking off so that the controlled equipment or distribution circuit may be worked on safely. Padlocking facilities can usually be arranged, but care must be taken to ensure that the facility cannot be defeated. Miniature circuit breakers used for isolation should provide clear indication that contacts are fully open.
3. Emergency stop controls must be placed adjacent to motors as with normal industrial practice. Shrouding should be provided in congested areas.
4. With the profusion of different junction, terminal and marshalling boxes around most installations, it is vital that good labelling is provided. Traffolyte types are usually acceptable, provided the lettering is dark on a light background. The reverse arrangements will lead to lettering quickly being obscured by dirt. Labels which depend only on glue for fixing are not recommended. All labels should be screwed or riveted. Engraved stainless steel labels are recommended for mechanically exposed positions.

7.14 Earthing

There are three different types of earthing system, as follows:

Electrical system earthing This is required for safe operation of the electrical system. It consists of a network of protective conductors. To ensure that fuses and protective relays operate effectively and that no dangerous voltage gradients exist, the resistance to earth of any part of this system should not exceed 1 ohm. Some oil companies require this value to be 0.5 ohm to provide a safety margin against the effects of corrosion and dirt. At main earth bars, facilities such as duplicate connections and removable links should be provided to allow for testing and maintenance to be carried out safely. Conductor sizes should have been already calculated during the design phase.

Static bonding This is required to prevent any arcs or sparks occurring between adjacent metal sections, such as gasketed pipe sections, cable tray and steel tanks. On machinery bolted to a steel deck, separate bonding connections are unlikely to be required. The resistance to earth and to adjacent metal parts should not exceed 10 ohms for any item.

Lightning protection   This form of protection is unlikely to be required offshore, but if it is found necessary according to the criteria in BS 6651 then the lightning rod to earth resistance should not exceed 7 ohms.

Earthing methods are detailed in the Institute of Petroleum Model Code of Safe Practice Part 1.

7.15 Installation in hazardous areas

7.15.1 Flameproof Ex'd' equipment

The requirements for flameproof equipment are discussed in detail in Chapter 8. The following criteria should be considered by the installer to ensure that certification is not invalidated and/or the risk of explosion increased:

1. Further entries should not be drilled out on flameproof enclosures. It is advisable to allow some margin in the number of entries specified to the manufacturer. Approved threaded plugs are available to plug unused entries. All unused entries must be plugged by such approved plugs.
2. Only tools suitable for use on flameproof equipment should be used.
3. Only flameproof cable glands should be used on flameproof enclosures.
4. The painting or obstruction of gaps between flanges must be avoided.
5. No flameproof equipment should be drilled, cut or welded.
6. Always check that all nuts, bolts, door hinges, isolators and interlocking devices are installed and are functioning as explained in the manufacturer's instructions.
7. All Ex'd' equipment flanges should be greased with a certified grease such as Chemodex copper grease.

It is often operator practice that flameproof equipment such as glands, luminaires and small motors are used throughout the installation in order to rationalize spares and simplify maintenance procedures. If this philosophy is in operation, the equipment sited in safe areas should be installed as if it were being installed in a hazardous area, with the correct seals, flange gaps etc. The reason for this is that the equipment will be marked Ex'd', and if hazardous zones are altered at some future date so that the equipment comes within such a zone, it will not be obvious that internal seals or other components are missing or not correctly installed.

7.15.2 Increased safety Ex'e' equipment

The requirements for increased safety equipment are discussed in detail in Chapter 8. The following criteria should be considered by the installer to ensure certification is not invalidated and/or the risk of explosion increased.

7.15.2.1 Terminals

1. The terminals should be installed on 32 mm carrier rail to DIN 46 277/1, each group being completed by an end section and sandwiched between two end support brackets.
2. Any cross-connected terminal assembly fitted with jumper bar must be mounted on the 32 mm carrier rail between insulating partitions.
3. Except when shown in a certificate as being internal wiring of the apparatus, not more than one single-strand or multiple-strand wire or cable core should be connected into either side of any terminal.
4. Leads connected to terminals should be insulated for the appropriate voltage, and this insulation must extend to within 1 mm of the metal of the terminal throat.
5. All terminal screws, whether used or not, must be tightened down.
6. When used in general purpose Ex'e' marshalling junction boxes, the circuits must be protected by close excess current protection which is designed to operate within four hours at 1.5 times the designed load current.
7. The creepage and clearance distances between the installed terminals and adjacent equipment, enclosure walls and covers must be in accordance with Tables 1, 2 and 3 of BS 5501: Part 6: 1977 (EN 50019).

7.15.2.2 Enclosure

1. Any cable gland or conduit entry into the enclosure must be capable of maintaining the degree of ingress protection IP54 and must be capable of passing the 7 N m impact test required by BS 4683 Part 4. If mineral insulated cable is used, the cable seal must be of a type having BASEEFA component approval for type 'e' applications.
2. The installer must ensure that the conductors are not cleated together in such a manner as to significantly increase the temperature of any individual conductor. A bunch of conductors at any point in the wiring loom should not contain any more conductors than are in the multicore cable or conduit from which the conductors originate.

Other design limitations are discussed in Chapter 8.

Chapter 8

Environmental topics

The environmental conditions for most electrical equipment offshore are generally more onerous than onshore. Environmental problems may be divided into three main categories:

(a)   those associated with sea and weather;
(b)   those associated with oil and gas;
(c)   those due to mechanical shock and vibration and the structural limitations of the installation.

8.1   Weather and sea protection

The salt-laden air and salt spray mists produced during bad weather have a searching corrosive effect on unprotected metal, as electrolytic cells are set up at these locations. Therefore any protective coatings used on metal equipment offshore should remain impermeable, and preferably contain a sacrificial metal such as zinc.

A variety of different enclosure and support structure materials have been tried offshore, with varying results, but experience suggests that the following materials are suitable:

Stainless steel   This tends to be an expensive solution, and needs to be carefully specified so as to avoid problems such as susceptibility to cracking. Stainless steel cable ladder and tray is more difficult to manufacture than that fabricated from the more common structural steels, and may be delivered with some deformity. This should be checked by normal quality control procedures, as even a small out-of-true error will make the installer's life very difficult. Stress corrosion in stainless steel takes place in the presence of chlorides, and related failures occur well below the normal tensile strength of the metal involved. The stresses during the deformation processes involved in the manufacture may remain locked up until 'season cracking' occurs in the presence of sodium chloride (salt spray). However, good quality low-temperature-annealed stainless steel products will provide long-term resistance to corrosion and are highly resistant to incidental damage.

Grey cast iron   In grey cast iron, most of the carbon is present in the form of graphite flakes, which make the material softer, more machinable and less brittle than white cast iron. As the name suggests, cast iron is very fluid when molten and is therefore suitable for the manufacture of intricate castings. Its main use offshore is in the construction of flameproof enclosures. Its resistance to corrosion appears to be quite variable. Where it is exposed to salt spray, for example in flameproof control stations on lower-deck handrails, a galvanized finish is advisable. The variation in the effects of corrosion is probably related to the method of producing cast iron, which involves the remelting of pig iron in cupolas. The qualities of the cast iron produced will depend on the selection of the pig iron, on the melting conditions in the cupola, and on special alloying additions.

Hot dipped galvanized steel   This material is by far the most common for use in cable support systems offshore. The heavy duty grade should provide a service life in excess of 20 years, particularly if Corten A steel is used.

Polycarbonate   This is a very tough plastic material used for junction box and similar electrical enclosures because of its corrosion-free property. It is particularly effective in areas close to the sea, where salt spray is common. It is resistant to mechanical damage and will deflect rather than break in most situations, but a heavy blow from a scaffolding pole, for example, is more likely to damage a polycarbonate enclosure than an equivalent steel one. Polycarbonate will tend to deform at elevated temperatures; it must therefore be shielded from the heat produced by flare stacks etc., and must not be used with equipment which has to operate during fire fighting operations.

Manganese bronze and gunmetal   These very heavy corrosion resistant metals are sometimes used with success for the casings of floodlights and similar exposed electrical equipment.

Welded and cast structural steel   Rotating machinery packages of all types are generally constructed of this material. It is not practicable to galvanize the whole package and therefore, as with the structural steel of the module or platform jacket, a suitable offshore paint system must be applied by the fabricator. The integrity of this paint system must be preserved during the equipment's transit, installation and commissioning, if the package is to be presented to the operator in good condition.

Glass fibre reinforced plastic   GRP has been used successfully offshore for a number of years. It may be used in electrical equipment in the form of small junction boxes and for cable ladder and tray. Although fire resistant, it will burn when subjected to a gas flame, but is more resistant to deformation, melting and fire than polycarbonate. Being strong and light and unaffected by sea water, it is useful particularly where installed in salt water spray conditions. However, as it will burn, long unbroken vertical runs should be avoided.

8.2   Enclosure ingress protection

All electrical equipment enclosures must be designed to prevent:

(a)   the inadvertent contact by personnel of live parts inside the enclosure, when these are accessed for maintenance etc.;

(b)   the ingress of dust and dirt;
(c)   the ingress of liquids, particularly conducting liquids such as sea water.

The degree of protection is specified in BS 5420 (and BS 4999 Part 105 for motors). The degree of ingress protection or IP rating is a two-digit number. The first digit refers to the degree of protection from foreign bodies and dust, and the second digit refers to the degree of protection from water. Some examples are given in Table 8.1. It should be noted that most hazardous area equipment enclosures must be capable of maintaining an ingress protection rating of IP54.

Table 8.1 Ingress protection rating examples

Code	First digit	Second digit
IP20	Contact with live or moving parts or ingress by objects with diameters of 12 mm and above	
IP24S	As above	Harmful effects of water splashed against the motor from any direction
IP44	Contact with live or moving parts by objects of 1 mm thickness Ingress of foreign bodies with diameters of 1 mm and above	Harmful effects of water splashed against the motor from any direction
IPW45S	Contact with live or moving parts by objects of 1 mm thickness Ingress of foreign bodies with diameters of 1 mm and above, and rain, snow and airborne particles to an amount inconsistent with correct operation	Harmful effects of water projected by nozzle against the motor from any direction
IP54	Complete protection against contact with live or moving parts inside the enclosure Protection against harmful deposits of dust	Water splashed against the motor from any direction shall have no harmful effect
IP55	The ingress of dust is not totally prevented, but dust cannot enter in an amount sufficient to interfere with the satisfactory operation of the machine	Water projected by a nozzle against the motor from any direction shall have no harmful effect
IP56		Motor protected against conditions on a ship's deck

8.3   Hazardous area applications

Much of the equipment on an offshore installation will be located in an area classified as potentially hazardous, because of the risk that flammable gases or vapours may be present in the atmosphere and could be ignited by equipment which creates electrical sparks or, during and perhaps for some time after operation, has an enclosure with a high enough surface temperature. This subject is covered exhaustively in a multitude of standards and codes of practice, many of which are listed in the Bibliography, and it is intended only to summarize the subject here. Those unfamiliar with the subject are recommended to study BS 5345 'Code of practice for the selection, installation and maintenance of electrical apparatus for use in potentially explosive atmospheres'.

An explosive atmosphere is one where a mixture of air and flammable substances in the form of gas, vapour or mist exists in such proportions that it can be exploded by excessive temperature, arcs or sparks.

The degree of danger varies with the probability of the presence of gas from location to location, and so hazardous areas are classified into three zones as follows:

Zone 0 (more than 100 hours per year) In which the explosive gas mixture is continuously present or present for long periods. The conditions in such zones are usually regarded as too dangerous for any electrical equipment to be located in.

Zone 1 (1 to 100 hours per year) In which an explosive gas/air mixture is likely to occur in normal operation.

Zone 2 (less than 1 hour per year but more than 1 hour per 100 years) In which an explosive gas/air mixture is not likely to occur in normal operation and, if it occurs, will exist for only a short time.

By implication, an area that is not classified as zone 0, 1 or 2 is deemed to be a non-hazardous area.

The numerical values for exposure time shown above are for guidance only.

8.3.1 Temperature considerations

8.3.1.1 Ignition temperature

The minimum temperature at which a gas, vapour or mist ignites spontaneously at atmospheric pressure is known as the ignition temperature. To avoid the risk of explosion, the surface temperature of the equipment must always remain below the ignition temperature of the explosive mixture.

Maximum permissible surface temperatures are classified as in Table 8.2. Classification applies to the equipment, not the gas; the actual classification designation can be taken as the next below the ignition temperature of the gas.

Table 8.2 Surface temperature classes

Temperature class	Maximum surface temperature (°C)
T1	450
T2	300
T3	200
T4	135
T5	100
T6	85

Motors to European standards conforming to EN 50 014 and 018 have temperatures 10°C lower than that shown in the table for classes T1 and T2, and 5°C lower for class T3 and below.

8.3.1.2 Flashpoint temperature

The flashpoint of a liquid or solid is the minimum temperature at which the vapour above the material just ignites by application of an external flame or spark, in standard test conditions.

The flashpoint gives a very useful indication as to how hazardous a material is, and is used when drawing up a schedule of hazardous sources for a particular installation. As discussed later in this chapter, equipment designed for use in hazardous areas should never allow sparks or flames to come in contact with the external environment, although in the case of Ex'd' equipment flames or sparks may occur within the enclosure. This is inherently the case with Ex'd' switchgear.

If the flashpoint of a substance is higher than 38°C, it is not normally regarded as a source of explosive hazard on North Sea installations. However, it should be noted that in hotter climates the storage area for the aviation kerosene (flashpoint 38°C) required for helicopters will have to be classified as hazardous, since there is a higher risk that ambient temperatures will exceed this temperature.

8.3.2 Explosionproof equipment groups

The maximum dimensions for flamepaths in Ex'd' equipment (see later), obtained from experiment, are used by the responsible standards authorities to subdivide flammable gases and vapours into three groups, A, B and C. The flamepath dimension is associated with the molecular size of the gas, the largest being in group A. Group C, which contains only hydrogen, is the most onerous as it requires the smallest flamepath cross-section and hence the finest machine tolerances.

Some examples of ignition temperatures and flashpoints are given in Table 8.3.

Table 8.3 Gas group examples

Material	Flashpoint (°C)	Ignition temperature (°C)	Temperature class	Gas group
Ethylene	gas	425	T2	IIB
Hydrogen	gas	560	T1	IIC
Methane	gas	538	T1	IIA
Kerosene	38	210	T3	IIA
Propane	−104	466	T1	IIA

8.3.3 Hazard source schedules

The process design must be scoured for every possible point where flammable substances are likely to leak. Only continuously welded pipes and vessels are not regarded as potential points of release. This task is best done by a team of engineers which includes a chemical or process engineer, an instrument/systems engineer and an electrical engineer. It is recommended that a hazop study be carried out concurrently with this task.

The resulting hazardous source schedule should give details of each source of hazard, its location, risk of occurrence, extent and type of hazardous zone produced, and remarks regarding any environmental aspects affecting the hazard such as dilution ventilation and airlocks. The schedule should also refer to a relevant and up-to-date hazardous area boundary drawing. If the installation already exists and is due to be modified or expanded, it will require to be thoroughly surveyed to ensure that the existing source schedules reflect the current situation. A typical hazardous source schedule is shown in Figure 8.1.

8.3.4 Defining boundaries

The extent of the hazardous zone round a source of release is usually determined in accordance with the British Institute of Petroleum Model Code of Safe Practice. This gives details illustrated by diagrams of typical situations with distances of the zone boundary from the point of release.

8.4 Ventilation

The degree of ventilation around the source of release will affect the classification of the hazard, unless one of the three following situations apply:

1. Open spaces with no structure or equipment restricting substantially free circulation of air, vertically and horizontally.
2. A module with a roof and not more than one side closed, free from obstruction to natural passage of air through it, vertically and horizontally.
3. Any enclosed or partly enclosed space provided with artificial ventilation to a degree equivalent to natural ventilation under low wind velocity conditions, and having adequate safeguards against failure of the ventilation equipment.

The number of air changes per hour required to meet the natural ventilation equivalent condition in the third case will vary to some extent, depending on the degree of hazard, but a figure of 40 changes per hour is quoted by underwriters. The positioning of fan inlets and exhaust outlets must be such that no unventilated pockets are left. The adequate safeguard is usually an alarm followed, after a suitable time delay, by the shutdown and venting to the flarestack of all process equipment within the module.

Process modules are usually ventilated such that the internal air pressure is slightly negative with respect to the outside, in order to retain any minor gas leaks. Safe area modules are kept at a slightly positive pressure in order to prevent the ingress of gas. The ventilation systems of different modules must be well segregated to prevent cross-contamination of leaks. Fire and explosion dampers are installed to automatically seal ventilation systems if fire or gas is detected. Ventilation intakes for both safe and hazardous

134

Equipment item		Flammable material	Operating conditions		Normal boiling point (degrees C)	Flash point (degrees C)	Auto ignition temperature (degrees C)	Vapour density (air = 1)	Source of release	Grade	Equipment temperature class	Equipment gas group	Zone category	Extent of zone (metres)	Remarks
Tag no.	Description		Temperature (degrees C)	Pressure bar Gauge											

Notes

Area:
Code:
Ventilation type:

Hazardous area classification schedule

Job no.

Drawing no.

| Revision & date | 3 | 2 | 1 | 0 |

Figure 8.1 Typical sheet from hazardous source schedule

areas should be located well within safe areas to minimize the risk of drawing gas into the ventilation system.

A common problem on older installations is the location of secondary pressurized control rooms in the middle of a hazardous module. A typical example of this is a gas compression module with its own control room. If the control room pressurization fails and remains failed for a timed period, the compression module should shut down and all non-certified electrical equipment in it should be isolated. Often, a loss of pressurization will only give an alarm; the decision to shut down the plant will be left with the operator, and prevention of gas ingress will depend on an airlock door system. Apart from the serious ignition hazard that all the live non-certified equipment poses in this situation, the control operator's only means of escape is via the compression module where the gas leak has occurred. Owing to the amount of work involved in providing automatic isolation, with shutdown contactors for all the electrical supplies, and separating the instrument and control functions so that the main installation control room is unaffected, it could well be more beneficial to relocate the compressor control room to a safe area or possibly include its functions in an extended central installation control room. The last option would be more in line with the Cullen Report, as it would remove an extra manned control room. Both relocation options eliminate the need to provide an escape route through the compression module protected to some degree from fire and explosion.

The ventilation of engine enclosures is discussed in Chapter 3.

A detailed discourse on the ventilation of offshore installations is beyond the scope of this book. However, the following is a list of basic criteria with which such ventilation systems should comply:

(a) the positive pressurization of non-hazardous areas with respect to adjacent hazardous areas or the external atmosphere;
(b) the containment, dilution and removal of potentially hazardous concentrations of explosive gaseous mixtures in hazardous modules, and the adequate segregation of hazardous area ventilation exhausts from ventilation and engine intakes;
(c) the provision of comfortable environmental conditions in accommodation and normally manned non-hazardous areas, and acceptable working conditions in normally unmanned areas;
(d) the provision of acceptable working conditions within hazardous modules;
(e) the isolation of individual areas and control of ventilation in emergency conditions, in accordance with the shutdown logic of the installation's ESD, fire, gas and alarm systems;
(f) the provision of combustion air for essential prime movers, ventilation air for escaping, fire fighting and rescue personnel, and purging air services required to operate effectively during an emergency.

To avoid depressurization and potential release of explosive mixtures, all exits to hazardous areas should be airlocked with self-closing doors. If single doors are used, a hazardous zone will extend outside the compartment.

Safe area control rooms may be located within hazardous modules provided they are kept pressurized with air obtained from a non-hazardous external area. Purging, alarm and timed shutdown facilities are required, similar to those for a pressurized motor. However, location of control rooms in safe areas is preferred for reasons given above.

8.5   Logic of area classification

Figure 8.2 shows a typical flow diagram for selection of hazardous zones. The flow diagram is used in conjunction with a questionnaire, such as that shown in Figure 8.3. The resulting zone boundaries are drawn on to a set of plans and elevations for the installation in order to produce the required hazardous area boundary drawings. Throughout the duration of a design project, regular meetings between discipline engineering heads and safety representatives are held to discuss and revise these drawings. Two typical hazardous area drawings are shown in Figure 8.4.

8.6   Selection of motors

It is common practice in offshore installations to require all motors for use in hazardous areas to be suitable for zone 1 areas, although in many cases the actual area classification might be zone 2. This allows areas to be reclassified from zone 2 to zone 1 without the need to replace motors; enables the safer transfer of motors from one hazardous area to another; and tends to rationalize spare parts holdings.

Where a motor is used as part of a variable speed drive system, the hazardous area certification will not be valid unless the motor has been tested for safe operation over the whole speed range through which it is required to operate. At low speeds the motor will have different heat dissipation characteristics and, if dependent for cooling on a rotor driven fan, will have a reduced cooling air flow rate. Therefore there is a risk that the enclosure temperature may exceed the required maximum surface temperature requirement.

It is now known that large motors whose casings are made up of bolted sections may suffer sparking across the section joints due to voltages produced by stray currents induced in the sections. To avoid risk of ignition, all such motors should now be fitted with copper braid bonding straps across each section joint in order to prevent any potential buildup across the joint.

In general, there are four methods of motor design to achieve suitability for use in hazardous areas, as follows.

8.6.1 Flameproof/explosionproof: Ex'd'

The principle of protection, as with the metal gauze on the Davy lamp, is to control the flow of and to cool any burning gas escaping from the motor casing, so that any gas that has escaped is no longer hot enough to ignite

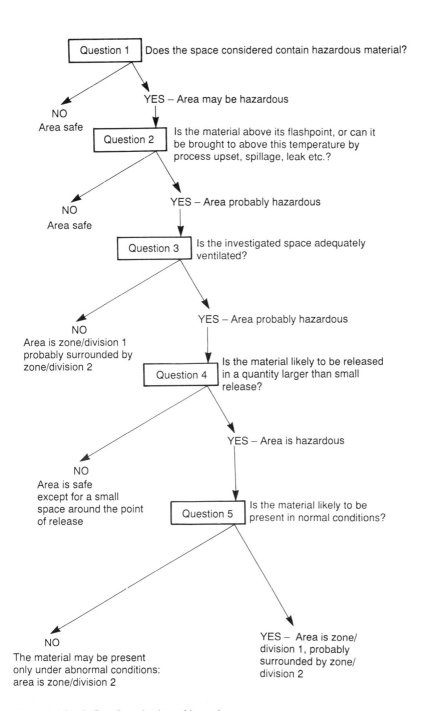

Figure 8.2 Logic flow for selection of hazardous zones

Hazard area classification sheet					Prepared by				
					Date				
					Issue	1	2	3	4
Client					Project no.				
Plant					Area		PDF no.		
Hazard point					Point no.		ELD no.		
Material	Vapour/gas	Liquid	Solid		Nature of hazard	Flash point	Material factor	Vapour density	Note reference
Component 1									
2									
3									
4									
Temperature of material					Flash point				

Density of the vapour/gas of the material relative to air at atmospheric pressure and ambient temperature, r =

1. Does the investigated space contain potentially hazardous material
   Otherwise than: (a) in pipes with no flanges, valves and fittings
   (b) stored in special containers approved for the material
   American   (c) in pipes with flanges, valves and fittings located in space which is **adequately ventilated**

   If Yes – Area may be hazardous   Yes ☐ No ☐
   If No – Area is safe

2. Is the material at a temperature higher than its flash point, or is there:
   (a) A source of heat at a temperature higher than the flash point of the material
   (b) An oxidising agent
   actually or potentially present within the distances from the point as indicated in the appropriate case in the examples of extent of hazard

   If Yes – Area probably hazardous   Yes ☐ No ☐
   If No – Area is safe

   Typical examples of extent of hazard areas

3. Is the investigated space **adequately ventilated**   Yes ☐ No ☐
   If No – Area is zone/division 1, probably surrounded by zone/division 2
   If Yes – Area probably hazardous

4. Is the hazardous material likely to be released in a quantity larger than **small release**
   If Yes – Area is hazardous   Yes ☐ No ☐
   If No – Area is safe except for small space around the point of release

5. Is the hazardous material **likely** to come into contact with atmospheric air in the **normal conditions**
   If Yes – Area is zone/division 1, probably surrounded by zone/division 2 Yes ☐ No ☐
   If No – Area is zone/division 2

6. Area and equipment classification
   British ☐   American ☐   Other (specify) ☐

   A. Maximum extent from the point of zone/division 1 area
   B. Maximum extent from the point of zone/division 2 area
   C. Ignition temperature of hazardous material

7. Notes

Figure 8.3 Questionnaire form to aid hazardous zone selection

external gas. The size of gap or flamepath necessary to sufficiently cool the ignited gas on its way out of the enclosure varies according to the gas or vapour involved. Gases and vapours are subdivided according to experimental data which has been established to determine the maximum experimental safe gap (MESG). In the case of metal-to-metal joints in a flameproof motor, for example that of the frame to the end shield, these will consist of a long metal spigot fitting into a long recess which will normally be clamped tightly to fixing bolts. A flamepath will always exist between the shaft and the motor interior.

For safety, all the flamepaths or gaps in the motor enclosure must never exceed mandatory dimensions, and the casing of the motor must be strong enough to withstand an explosion caused by the ignition of the maximum free volume of air/gas mixture it can contain, with all flamepaths at minimum production values. It should be noted that this could be exceeded in very cold regions (polar latitudes) owing to the increased density of the gas, or at the elevated pressures used in diving. This type of protection is common on low-voltage motors up to a 315 frame size; on larger sizes, severe cost and weight penalties are likely owing to the cast method of construction.

8.6.2 Increased safety: Ex'e'

This type of protection relies on the reduction to negligible of the risk of an explosion by careful design of the motor and its control gear in order to eliminate any potential sparking which may come in contact with an air/gas flammable mixture, or an excessive temperature anywhere within or on the surface of the machine. Design criteria for this type of motor are given in BS 5501 and BS 5000.

This type of protection is not used in North American equipment, and the standards have therefore been written with European voltages only in mind, quoting a maximum nominal system voltage of 11 kV. Strictly speaking, equipment made for the North American standard voltage of 13.8 kV cannot comply with these standards. However, it is usual under these circumstances for such motors to be tested and then issued with a certificate of inspection rather than a certificate of authority. Acceptance of this method of protection must therefore be obtained from the platform certifying body and underwriters before any decision to use such a motor is made.

Ex'e' motors offer increasing savings in weight and bulk as ratings increase. Beyond the 315 frame size, the only economic alternative is the Ex'p' type of protection.

8.6.3 Non-sparking: Ex'N'

The design concept for Ex'N' machines is similar to Ex'e', but among other minor differences lacks several important features. First, the Ex'N' certification does not include the starting condition. Unlike an Ex'e' machine, the surface temperature of the casing is allowed to exceed the maximum permitted by the temperature class during starting. Secondly, no special

140

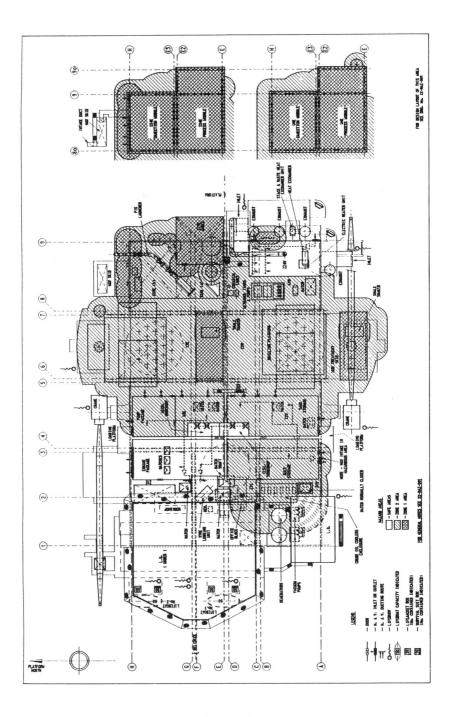

Figure 8.4 Typical hazardous area boundary drawings

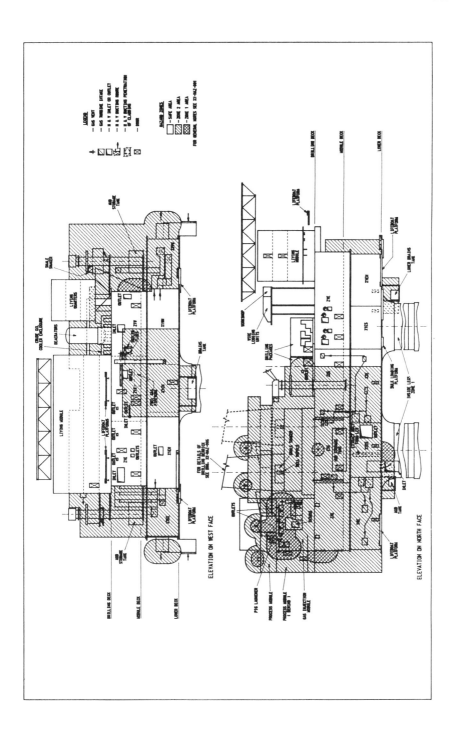

overload protection is called for. Because of these limitations, Ex'N' machines are restricted to use in zone 2 or safe areas only.

Problems have recently been experienced with large Ex'e' and Ex'N' motors operating at voltages over 3.3 kV where the motor casing itself has exploded. Research into the problem is still in progress, particularly as both the source of ignition and the source of gas have been puzzling the engineers concerned. The ignition source appears to be associated with partial discharges across the end windings in machines exposed to heavy salt or other contamination. Several operating conditions have been identified where gas could migrate into the motor casing to provide an explosive concentration. First, where a gas compressor or crude oil pump shares a common lube oil system with its drive motor, gas can become entrained in the lube oil at the process end, and, on reaching the motor bearing where the pressure is lower, leave the lube oil. Therefore common lube oil systems should not be used. Secondly, minor gas leaks often occur on gas compressor packages, and the cooling of the motor casing after a shutdown may draw gas in, so that a hazardous concentration exists within the motor when it is restarted. Apart from the more drastic option of replacement or recertification as a pressurized unit, the motor could be fitted with a gas sampling unit which will alarm on the presence of gas within the motor casing, a nitrogen prestart purging kit, or both.

8.6.4 Pressurized: Ex'p'

A pressurized motor relies on maintaining all internal parts of the motor enclosure at a greater pressure than atmospheric, in order to prevent the ingress of any hazardous gases that may be present in the vicinity of the motor. No special precautions need to be taken with winding temperatures, but the motor interior, especially the stator windings, must be designed to allow a purging air flow to clear any pockets of flammable gas, within the required purging time. Be warned that this may not be the case with the standard industrial version of the machine.

To achieve pressurization, an external source of dry air must be provided from a safe area. Although this type of motor is inherently cheaper than an Ex'e' equivalent, the cost saving may be outweighed by the extra costs for the air supplies, pipework and control systems necessary. Another possible cause of delay and expense is the need for special certification to be obtained from an authorized certifying authority such as BASEEFA or INIEX.

8.7   Selection of other equipment

Much of the explanation above regarding hazardous area motors may be applied to other hazardous area equipment. Switchgear, being inherently sparking equipment, must be enclosed in an explosionproof or pressurized enclosure, as it cannot meet the criteria for increased safety Ex'e' certification.

Enclosures used for housing cable terminals may be of explosionproof or increased safety design. Common types of enclosure and limitations affecting their use and those for cable glands and transits are discussed in Chapter 7.

Hazardous area lighting is discussed in Chapter 10.

8.8   Structural considerations

8.8.1 Weight control

As in ship construction, the recording and control of weight are of vital importance to the structural integrity of the installation and the safety and survival of those on board. As most equipment is shipped offshore in the form of prefabricated, precommissioned modules, weight control must start at the fabricator's yard and in the manufacturer's works. The selection of equipment may be restricted by maximum weight allowances. The power-to-weight ratio of a particular prime mover may be critical if the projected electrical load is to be met without the weight limit of the generator module being exceeded.

The weight of cabling and support systems, although more distributed, makes a significant contribution to the total, and must be included in the weight control schedule. This schedule is usually compiled and updated regularly by structural engineers, throughout the duration of a design contract.

The total weight of every module may be restricted, not just by the load bearing capabilities of the offshore structure, but also by the capabilities of offshore craneage available. The lifting capability of an offshore crane will vary depending on the operating radius of the jib required and the sea state, particularly the wave height, in which the crane is required to operate. Therefore if a module weight is known to be critical, switchgear and other heavy equipment should be installed using bolted brackets rather than welding, as it may be necessary depending on the sea state etc. to temporarily remove the equipment and reinstall it offshore.

8.8.2 Shock and vibration

With platform structures installed in sea depths of 150 metres or more, rotating machinery rated at 5 MW or more may be operating near the top of a structure over 200 metres high. The structure may resonate with vibrations produced by the machinery and, particularly if the machinery skid floor is cantilevered out beyond the main structure, a catastrophic failure may occur. To ensure that this is not the case, vibration analyses need to be carried out to establish structural resonance frequencies. If these are close to those of the machinery, modifications will need to be carried out in order to change the offending frequency to an acceptable value. At such heights, the swaying of the platform and the shock from drilling activities etc. may produce damaging torques on the bearings and shafts of large motors and generators if these are not catered for in the machine design.

8.8.3 Location of engine intakes and exhausts

An offshore production platform is a compact three-dimensional arrangement of modules. Some modules contain prime movers; others, such as separator and gas compression areas, require to exhaust ventilation air from hazardous areas; and all must take in fresh, uncontaminated air from the surrounding atmosphere. The prevention of cross-contamination between exhausts and intakes in all wind and weather conditions can be extremely difficult, and in some weather conditions it may be necessary to accept a small dropoff in gas turbine performance due to the thermal contamination of other exhausts. Water curtains around turbine exhaust ducts are sometimes used to good effect in reducing contamination, by taking exhaust gas downwards, away from intakes. Prime movers for emergency equipment such as emergency generators and fire pump alternators are not so critical, because testing can be carried out in a favourable wind which blows the exhaust away from ventilation intakes. With emergency equipment, it is more important that aspects of emergency scenarios such as gas cloud boundaries and likely flame passages are given consideration.

8.8.4 Mechanical protection

On any platform, routine production maintenance and drilling operations demand the movement of equipment, containers, scaffolding poles, drill-pipe etc. from the laydown area, at which it was received from the supply vessel, to its point of use, and then possibly the reverse journey back to the laydown areas prior to transfer back to the supply vessel. During these movements there is a risk that the item will be dropped on to, or swung into, some exposed piece of electrical equipment. Worse still, an item may be dropped by a crane so that it pierces the roof or wall of an operating module. To avoid such occurrences the following should be considered.

First, exposed items of electrical equipment located in busy thoroughfares such as walkways, drilldecks and laydown areas should be of steel construction or provided with steel impact protection of some form. No equipment should obstruct walkways. Overhead equipment such as luminaires and cable tray should not project lower than 2 metres (safety helmets add about 100 mm to a person's height).

Secondly, the loci of crane loads should be carefully studied so that areas where the risk of dropped loads is high can be avoided as prospective critical equipment locations. If high-risk areas are unavoidable, then adequate mechanical protection will need to be provided, allowing for the shape and kinetic energy of potential dropped loads.

8.9 Noise control

It is practically possible to run a platform unmanned and to shut down only for planned maintenance periods, which would then be the only time when the maintenance crew would be present. However, equipment reliabilities demand that unplanned outages be attended to, so noise levels must be limited throughout any installation if it is to be operated and manned at the same time. Sound levels of under 85 dB(A) should be the aim of any design, with much lower levels near accommodation areas.

8.9.1 Prime movers

Generator set procurement specifications must detail the maximum sound power level acceptable, with frequency spectra if possible. Equipment tests will need to be witnessed to ensure that the specifications are complied with. However, three main methods are available to the offshore design engineer to reduce the effects of prime mover noise offshore:

1. A close fitting acoustic hood can be installed over the engine, usually sized to fit round the skid on gas turbines. The enclosure sound insulation material must not absorb fuel, lubricant or coolant, as the enclosure itself must be designed to contain and extinguish engine fires with the aid of a halon or similar gas discharge.
2. It is difficult to enclose reciprocating engines in this way, without impeding access for maintenance. Therefore the module housing the engine and generator may be soundproofed. If acoustic dampers are fitted to ventilation intakes and exhausts, this has the advantage that generator and radiator fan noise are also controlled.
3. Modules containing normally running prime movers should be located as far away from continuously manned areas as possible, although this distance is likely to be constrained by other limitations such as hazardous areas and air intake and exhaust positioning.

In practice, a combination of all three methods will be used.

8.9.2 Motors

Unless the predominant noise is from the driven equipment, sound insulation of the module is not usually provided, since it should be possible for the motor manufacturer to control noise in the machine concerned. The specifier, however, may assist by avoiding the use of two-pole motors wherever practicable.

The sources of induction motor noise are as follows:

(a) low-frequency magnetic noise from stator core laminations (100 Hz);
(b) fan and air flow noise (400–2000 Hz);
(c) core vibration noise (1000–2000 Hz);
(d) slot tooth vibration noise (2500–4000 Hz);
(e) bearing noise (6000–8000 Hz).

The magnetic noise can never be totally eliminated, since it is dependent on a large number of variables, which the machine designer must consider in order to trade off minimum noise against optimum performance and efficiency. Fan noise can be reduced by fitting a smaller fan on an oversized motor and also by employing closed air circuit (CACA) cooling with an air-to-air heat exchanger. Silencing material may be fitted to the cooling air circuit, but this needs to be of a type suitable for the environment in that it should not absorb dirt, oil and sea spray as well as it does sound. The ingress of dirt etc. is much less with CACA machines.

If a cooling water supply is available, then a substantial noise reduction may be obtained by replacing the air-to-air heat exchanger with an air-to-water type. Be warned, however, that the availability of the machine will then be dependent on the availability of the cooling water supply, which may itself be dependent on the satisfactory operation of a number of sea water lift and circulating pumps.

Chapter 9

Power system disturbances: prediction and protection

As discussed in earlier chapters, the isolated location, onerous operating conditions and harsh environment offshore make it particularly important that provisions are made in the design in order to protect electrical equipment from faults in the system, and to remove faulty equipment from the system quickly and safely. Alternators must be protected from prime mover and fuel system faults. Motors must be protected from faults in driven equipment. In this chapter, common protection relay applications are discussed and worked examples are provided for power system relay configurations likely to be met in offshore installations.

9.1 Alternator faults and protection devices

9.1.1 Neutral earthing and earth faults

Medium-voltage generators are normally resistance earthed in order to limit fault currents. Stator windings should be designed to minimize third-harmonic circulating currents, so that generators may be paralleled without extra weight and space being taken up with earthing circuit breakers. The short time ratings of any earthing resistors must be borne in mind when calculating earth fault relay settings. A current sensing relay of either inverse or definite time characteristic may be used for unrestricted earth fault protection. Alternatively, an attracted armature relay may be used in conjunction with a time delay relay. In either case, the time delay must be within the time rating of the earthing resistor and the relay must coordinate with those downstream. The term 'backup' earth fault protection is sometimes used to describe this scheme, as it supplements the generator differential protection. However, unrestricted earth fault protection will not detect phase-to-phase faults, and therefore the term 'backup' or 'standby' leads to some confusion and is not recommended.

The neutral earthing conductor arrangement must be rated for the prospective earth fault current flows available from the generator. If the generator is a low-voltage machine with bolted earth connection, this current is likely to be high enough to affect cable conductor sizes. With directly earthed low-voltage machines, the high currents are used to operate fuses and miniature circuit breakers (MCBs) usually found in low-voltage distribution systems.

With larger medium-voltage machines the damage caused to laminations is limited by the insertion of the neutral resistance, and more sensitive forms or protection may be used in the medium-voltage distribution system.

9.1.2 Overload protection

The larger offshore alternators will not have overload protection as such, but resistance temperature devices (RTDs) are buried in the stator windings, and will give alarm and trip signals at temperatures where overloading or other abnormal conditions such as excitation faults have occurred. The temperature protection will avoid unnecessary shortening of insulation life.

With small emergency generators, the associated circuit breaker should be provided with a thermal element.

9.1.3 Overcurrent protection

To protect the generator from downstream faults, overcurrent relays of either the inverse definite minimum time (IDMT) or the definite time type are required.

The protection relay must provide sufficient time for downstream protection relays to operate to clear the fault, but this time must be within the time rating of the alternator. Larger generators are usually provided with permanent magnet pilot exciters to maintain a fault current of around three times rated current continuously after the transient short-circuit current has decayed away. This can be maintained usually for an absolute maximum of 10 seconds before the generator overheats.

However, it should not be necessary to approach even 5 seconds in order to operate downstream protection, and the damage due to arcing faults if such a large amount of fault energy is let through can be extensive. It is therefore advisable to keep fault times to a minimum by good relay coordination.

With small self-excited generators it is important to specify a fault current maintenance kit consisting of a voltage sensitive relay and compounding current transformers (CTs) designed to maintain excitation current when the output voltage collapses owing to a fault.

If reasonable coordination cannot be obtained using normal IDMT or definite time relays, voltage controlled or voltage restraint relays can be used to take advantage of closeup fault voltage collapse. An example of this is given in Section 9.9.

9.1.4 Phase and interturn faults

Stator and interturn faults are relatively uncommon, but are more likely to occur at the ends of the windings or in the terminal box. Insulation failures between phases may be sensed by differential protection. Both types of fault will cause heating and possibly a fire, which should be detected by suitable smoke detection in the acoustic hood, if not by the RTDs or the differential protection.

9.1.5 Winding protection

Restricted earth fault (if the individual phases are not accessible at the neutral end) or phase and earth fault circulating current protection, is usually provided on machines of rating 500 kW and larger, although this threshold will vary with the degree of criticality of the supply. A stabilizing resistor must be provided in series with the relay to prevent CT saturation effects, causing the relay to operate in through fault conditions. The method of calculating the resistance value is given in Section 9.8.

9.1.6 Over/undervoltage protection

This facility is often provided as part of the generator's automatic voltage regulator (AVR), but if this is not the case it must be included in the generator control panel or switchboard. If fed from a separate voltage transformer (VT), it may duplicate the AVR facility, so that failure of one set of VT fuses cannot lead to loss of voltage control and/or monitoring.

9.1.7 Over/underfrequency protection

This facility is often provided by the prime mover governor, but as it is more critical on a generator isolated from a grid system, it is often duplicated electrically. This ensures that the circuit breaker trips before damage is done to motors or driven machinery, even if the prime mover overruns owing to gas ingestion at the air intake (see Chapter 3).

9.1.8 Unbalanced loading and negative phase sequence protection

If the platform load is unbalanced, there will be a negative phase sequence component in the generator load current which, if excessive and over a long period, will cause overheating of the rotor. On the larger offshore generator packages, it is therefore necessary to fit negative phase sequence protection which, after a timed period corresponding to the thermal characteristic of the rotor, will trip the generator. As the source of negative sequence current is external to the generators and will lead to the tripping of all the generators if not removed, it is normal to provide an alarm at the onset of the problem in order that operators may have time to find the offending load before a production shutdown occurs. A suitable setting value should be sought from the generator manufacturers in each case, but a value no greater than 30% is recommended.

9.1.9 Rotor faults

The exciter output current on brushless machines is monitored and, should a rotating diode fail, an alarm will be annunciated on the generator control panel. This facility is usually included in the AVR circuitry and therefore it is not likely to be included in the discrete protection relay suite.

9.1.10 Field faults and asynchronous operation

If the generator field current fails and the generator is running as the sole supplier of power on the installation, the set will trip on undervoltage, causing a blackout until the emergency generator starts. However, if the generator is running in parallel with a second machine, it would continue to generate power as an induction generator, whilst demanding a heavy reactive power flow from the machine in parallel with it. Both machines will tend to heat up, and in some cases it may appear as if the healthy machine is the offender.

A field failure relay of the 'Mho' impedance type is normally used to protect generators from this condition. The generator reactance for a machine where the excitation has failed is not fixed but describes a circular locus, and so the relay characteristic should be set to enclose this locus as fully as possible.

9.1.11 Package control and supervision: master trip and lockout relays

Individual relay protection elements are usually provided without any form of lockout facility, although some means of electronically or mechanically (i.e. flag) recording relay operation is often included. Where a switchboard is fed by a number of generators, the operational logic of the system must be carefully studied to ensure every mode of operation possible is reliable and safe. It is usual to provide a master trip relay for each generator, which will operate from a signal derived from any of the fault sensing relays and then lock the generator out of service. Without the master trip relay, indication of the fault on the control panel could well be lost, and there would be nothing to prevent an operator attempting to restart the machine before the fault is cleared.

If failure of one generator leads to automatic starting of another, it is important that the control logic differentiates between generator failure and uncleared switchboard faults, in order to avoid the second generator circuit breaker closing on to the same (switchboard) fault. The generator circuit breaker tripping coil and its circuit should be monitored by a trip supervision relay, which provides an alarm on failure of coil continuity or loss of tripping supply. Because of the risk of generator, switchboard or transformer fires due to sustained fault conditions if a generator circuit breaker fails to trip, and their possibly serious consequences offshore, it is advisable to fit (to main generator circuit breakers at least) trip supervision relays which continuously monitor both tripping supply and trip circuit continuity.

9.2 Transformer faults and protection devices

9.2.1 Transformer faults

9.2.1.1 Earth faults

The magnitude of any fault current flowing in a transformer will be a function of the winding arrangement (and hence leakage reactance) and the type of earthing (solid or impedance). Most offshore distribution transformers, however, are of the delta star (usually Dy11) type, the only exceptions normally being those used for supplying the drilling SCRs. With the usual solidly earthed star point arrangement, the fault current will be highest when the fault is near the neutral. There is no zero-sequence path through such a transformer, and therefore coordination is not required with any earth fault protection on the primary side.

The magnitude of any earth fault on the delta primary winding will be governed by the type of earthing in that part of the system. If it is fed from the main switchboard, it will probably be limited by the generator earthing resistors.

9.2.1.2 Phase-to-phase faults
Faults between phases on offshore transformers are relatively rare, especially with the newer sealed or encapsulated types of transformer. Such a fault is likely to produce a substantial current, large enough to be detected quickly by upstream overcurrent protection.

9.2.1.3 Core and interturn faults
Although interturn faults are unlikely on offshore distribution transformers because of the relatively low voltages, the fire hazard is such that their probability cannot be ignored. A fault in a few turns will cause a high current to flow in the short-circuited loop and produce a dangerous local hot spot. A conducting path through core laminations will also cause severe local heating. It is difficult to detect such a fault in a resin encapsulated transformer.

Offshore distribution transformers are less likely to suffer line surges or steep fronted impulse voltages since cabling is relatively short and lightning strikes are rare. However, moisture ingress is more likely in the offshore environment, and mechanical vibration levels are higher, making chafing and cracking of insulation more likely, so the possibility of interturn faults cannot be ignored.

If the transformer is housed in a tank containing insulating oil, there is always a danger that the tank will corrode in the salt-laden atmosphere, or will be damaged by crane operations or the like.

Offshore transformers are as likely to suffer from abnormal system operation as their onshore counterparts. The maintenance of system voltage and frequency may be dependent for long periods on the satisfactory operation of one generator package. The governor or AVR of this package may be set low or high, or may be subject to drift within the limits of the voltage and frequency protection. Therefore transformers may be subject to variations in voltage and frequency. High voltage and low frequency combined may cause shifting of flux to structural parts of the transformer, which will heat up and destroy insulation. The transformer must also be protected from overloads and overcurrents due to downstream faults.

9.2.2 Magnetizing inrush

Magnetizing inrush is a normal healthy but transient condition associated with the establishment of linking flux between the windings. However, if it is forgotten when setting upstream overcurrent protection relays, it may cause nuisance tripping. Values quoted by transformer manufacturers are in the region of 12 times full load for 15 milliseconds.

9.2.3 Overcurrent protection

A common arrangement offshore is to protect the primary winding of the transformers between the main (MV) and the production (LV) switchboards with IDMT overcurrent relays graded with the main generator overcurrent protection. The relay may have an earth fault element, able to detect earth faults in the primary winding only.

9.2.4 Restricted earth fault protection

With the restricted earth fault scheme, the residual current obtained from a CT in each line is balanced against the current from a CT in the neutral. The neutral is usually solidly earthed and therefore a substantial fault current will be produced even from a fault at the last turn of the winding (i.e. closest to the neutral). As faults are only detected between the line CTs and the neutral CT, only the star secondary is protected, using a sensitive instantaneous relay.

9.2.5 Differential protection

Differential protection is rarely used on transformers offshore. However, it does have the advantage over restricted earth fault protection in that both primary and secondary windings are protected from both earth and phase-to-phase faults. The relay will only operate for faults appearing in the protection zone between the sets of sensing CTs at the primary and secondary terminals. Under normal load conditions the CT secondary currents are equal and no current flows in the relay operating coil. If these currents become unequal owing to a fault current being sensed in the transformer windings, the resulting current energizes the operating coil. The relay contacts close when the ratio of this differential current to the through current exceeds the slope of the relay operating characteristic. This characteristic may be altered by adjusting the turns ratio of the operating and restraint coils.

The bias slope is chosen so that the relay remains insensitive to the spill current produced by through currents flowing to loads or faults external to the transformer. Spill currents occur due to errors that arise in the matching of current transformers and other components in the relay circuit. In addition, a fairly high bias slope is required to prevent maloperation by CT differential currents arising from:

(a)   tap changing on transformers giving CT mismatch;
(b)   different CT ratios and hence saturation levels giving differential currents under through fault conditions;
(c)   magnetizing inrush giving secondary currents in one set of CTs only.

To prevent maloperation by magnetizing inrush, the relay operation is delayed by a selected time delay which allows the initial current peaks to decay to below the relay setting value determined by the percentage bias.

9.2.6 Oil and gas operated devices

Pressure or flow switches of various kinds may be fitted to oil filled transformers, in order to detect faults in the windings which give rise to sudden pressure increases in the tank or sudden flow to the conservator (i.e. Buchholz devices). Conservators are rarely fitted offshore, since sealed silicon oil designs are now preferred in order to reduce fire risk; thus a rate-of-rise pressure sensing device is usually fitted on the tank. Protection of this kind, which minimizes the risk of transformer fires, should be considered for all but the smallest of offshore distribution transformers because of the serious consequences such a fire would have.

9.2.7 Parallel transformers and intertripping

9.2.7.1 Momentary paralleling schemes

In order to allow for transformer and associated equipment maintenance and for unplanned outages, it is usual to provide two transformers in parallel as feeders to the production LV switchboard. In most systems, each transformer is rated for the full production switchboard load, but the production switchboard is only fault rated for operation with one transformer connected. In such a case, interlocking and paralleling facilities are provided which only allow momentary paralleling as the load is transferred from one transformer to the other. The risk that a fault occurs during the changeover period is usually very small and therefore acceptable.

Where the LV switchboard has a bus section switch, a third operating alternative is available where each section is fed from one transformer. This has the benefit that one transformer fault will only require the reinstatement of one LV bus section and will not disrupt supplies to the bus section fed by the healthy transformer.

9.2.7.2 Intertripping

A faulty transformer must be isolated by the circuit breakers on both the primary and secondary sides, wherever the fault is sensed, for two reasons. First, a fault sensed by protection only on the secondary circuit breaker would leave the transformer energized in a dangerous state. Secondly, a fault sensed by protection only on the primary circuit breaker could allow the transformer to be back fed from the secondary circuit, again leaving the transformer energized in a dangerous state.

A scheme where the primary and secondary circuit breaker trip circuits are linked is known as intertripping. Momentary paralleling and intertripping schemes are often combined, and the scheme may have additional facilities, such as those required for the emergency back feeding of a main switchboard air compressor when main generation is not available.

9.3 Motor faults and protection

The various types of motor used offshore are discussed in Chapter 4. This section describes motor protection devices and their application offshore.

9.3.1 Drive system faults

In general there are four main categories of drive system fault which need to be protected against: electrical faults and three types of mechanical fault. These are as follows.

9.3.1.1 Motor winding electrical faults
Such faults are usually catered for by means of fuses or circuit breakers fitted with instantaneous overcurrent relays. Sensitive earth fault relays set at about 20% of full load current are normally provided on motor starters rated at more than about 40 kW.

9.3.1.2 Bearing failures
Without some form of vibration monitoring, it is not very practicable to detect the incipient failure of a ball or roller bearing and to shut down the machine before the bearing disintegrates. However, monitoring devices are being developed which detect the effect of various forms of abnormal vibration on a machine's magnetic signature. This information is then processed to provide various alarms and trips.

The incipient failure of a sleeve bearing can usually be detected by an RTD or similar device which detects the rise in bearing temperature before seizure or disintegration.

9.3.1.3 Abnormalities in the driven machinery
The control logic of the driven machinery package should prevent the occurrence of severe changes of torque, negative torques, or reverse running before the motor is energized. An example of such a fault would be if an extra water injection pump was being brought into service, and sea water, pressurized by the running pumps, was allowed to flow backwards through the pump being started, thus driving the pump and the motor backwards. The extra time at elevated current required by the motor to decelerate before accelerating in the forward direction may be sufficient to overheat the machine. It may be necessary to fit a tachometric device to the machine such that reverse running will result in an 'inhibit start' signal. Although valve actuator interlocking logic should be designed to prevent the water valves being opened in the wrong sequence, it is worth checking the driven equipment control scheme to ensure that this or any other likely cause of motor failure is catered for by one or more protective measures.

High winds blowing directly on to ventilation fans, causing them to run backwards at high speed, create a similar problem. In this case, non-return dampers or brakes which release when the motor is started can be used.

9.3.1.4 Abnormalities in the supply system
The supply system depends on a limited number of platform generators which are, in turn, dependent for their fuel, cooling water etc. on the correct operation of other equipment on the installation. Therefore there is a significant probability that frequency and voltage fluctuations will occur of sufficient magnitude to affect the running of motors. With low-frequency problems, stalling protection will adequately deal with the situation. High-frequency problems should be catered for by overload protection devices.

However, with large reciprocating compressors it may be advisable to fit timed overfrequency and/or underfrequency trips to avoid running the machine for long periods at a frequency close to tolerance limits. Generator under/overfrequency trip settings should be a compromise between taking the most sensitive plant into account, and introducing the likelihood of generator nuisance tripping.

Running at abnormally low system voltage will cause motors to draw larger currents and increase heating effects in motor windings and supply cables. A lower system voltage will also increase the likelihood of stalling, as the motor torque is proportional to the square of the voltage at a given speed.

An abnormally high voltage will reduce heating effects in windings and supply cables, and give faster starting times owing to the increased torque. Because of this, it is the practice on some installations to run the system at a few per cent above the nominal voltage. Unfortunately, prospective fault currents increase in proportion to the square of the voltage. Therefore, if the system is to be normally run at a voltage several per cent above nominal, the prospective fault currents should be recalculated to reflect this, and switchgear fault ratings checked accordingly.

9.3.2 Overload protection

The heating effects due to motor overloads can be simulated to some degree by a bimetal element or an electronic thermal replica. Both the stator windings and the rotor will begin to heat up at the onset of an overload. At ratings below about 20 kW, depending on the design and type of machine, the stator windings will tend to suffer thermal damage in a shorter time than the rotor, in which case the motor is termed 'stator critical'. Stator critical motors are easier to protect than rotor critical types since the stator windings are more accessible than the rotor, and can be fitted with temperature sensors if the thermal overload device is felt to be inadequate. If the motor is driven by a variable speed drive, where it is likely to run for some time at low speed, the thermal effects of reduced ventilation must be taken into account in the choice of motor and the setting of the overload. The variable speed drive unit is usually fitted with sophisticated overload protection as a standard feature.

Electronic replica protection relays are normally reserved for the protection of larger machines. However, a drive of only a few tens of kilowatts which has a particularly critical function, or is located at some part of the platform where any form of overheating could not be tolerated, for example, an Ex'e' motor in a hazardous area, may also require such a device to be fitted.

9.3.3 Stalling protection

When the elevated current drawn by a motor during starting is prolonged, for example because of a mechanical fault in the motor or driven machinery, then the motor will very rapidly overheat unless it is quickly disconnected from the supply.

If the probability of a motor stall is considered significant for a particular drive package, a stalling relay should be fitted. This device is a thermal element similar to the overload element, but designed to respond to the higher short-term starting currents, and to trip if the starting current continues for more than 5 to 10 seconds beyond the normal starting time.

Normal starting times vary depending on the characteristic of the driven equipment. As they increase, discrimination between starting and stalling conditions becomes more and more difficult, until the point where it becomes necessary to fit a tachometric device to the drive shaft in order to detect the absence of rotation or the failure to achieve normal running speed.

9.3.4 Phase imbalance protection

The loss of one phase of the supply or unbalanced supply voltages will have two effects. First, it will create a negative phase sequence current which will cause additional rotor heating. Secondly, it will cause excessive heat in the stator windings if the current in one or more phases exceeds the normal full load value.

The heating effect due to the negative sequence current can pose a greater threat than that due to the unbalanced phase currents, and therefore it is advisable, especially with larger machines, to provide a relay with negative phase sequence detection. The operation of the phase imbalance facility on the P&B Golds relay is described in the next section.

9.3.5 Conventional relay types

The most common of the conventional thermal overcurrent relays offshore is the P&B Golds relay.

The relay consists of three tapped saturating core current transformers which are fed from the main protection CTs. The secondary of each saturating core CT feeds a helical heating coil element which operates its own moving contact. The construction of the basic thermal overload element is shown in Figure 9.1. The centre phase heater is used to provide

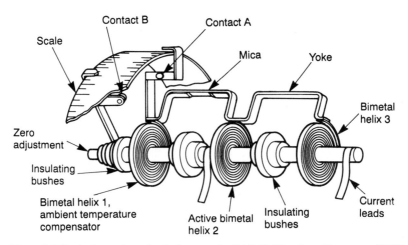

Figure 9.1 Basic thermal overload element of a P&B Golds relay. (Courtesy of P&B Engineering Ltd.)

'% running load' indication; overload trip when the moving contact touches the '% load to trip' contact; and unbalanced load trip when contact with either of the outer phase contacts is made. Under healthy operating conditions, the out-of-balance contact cradle is horizontal so that no contacts are touching. However, if the motor load becomes unbalanced or single phasing occurs, the differential heating effect will cause the cradle to tilt over until the contacts touch and the motor contactor is tripped.

9.3.6 Static and microprocessor based relay types

Examples of such devices available in the early 1990s are:

P&B Micro-Golds	P&B Engineering, Crawley, West Sussex
Sprecher & Schuh CET3	Sprecher & Schuh, Thame, Oxfordshire
Multilin model 169	Multilin, Markam, Ontario, Canada
GEC Alsthom Measurements MCHG/N	GEC Alsthom Measurements, Stafford
Krauss & Naimer IPD	UK Solenoid, Newbury, Berkshire

Figure 9.2 shows the schematic of a typical electronic thermal replica relay.

9.3.7 Additional protection for synchronous motors

The following additional protection devices may be required for synchronous motors:

Pullout protection If the compression torque for a reciprocating gas reinjection compressor, for example, exceeds the motor pullout torque, the motor comes out of synchronism and stalls. To prevent the motor continuing to draw a heavy stator current, both the motor supply and the field supply are tripped.

Damper winding protection The damper windings are often designed to allow the machine to be run up as an induction motor. However, if an excessive number of starts are attempted in succession, the windings will overheat. A thermal relay should be provided, which will trip the motor supply before any damage is done.

Reverse power If there is a disturbance in the supply system or the supply fails, the machine will generate using the inertia of the drive train for power. In this situation, the motor should be disconnected if the stability of the system is not regained after a short (timed) interval.

9.4 Busbar protection

9.4.1 Busbar faults

Busbar faults in medium- and low-voltage metal clad indoor switchgear are very rare in onshore installations. However, several such incidents have occurred offshore. In one instance, where the risk of busbar faults was

158

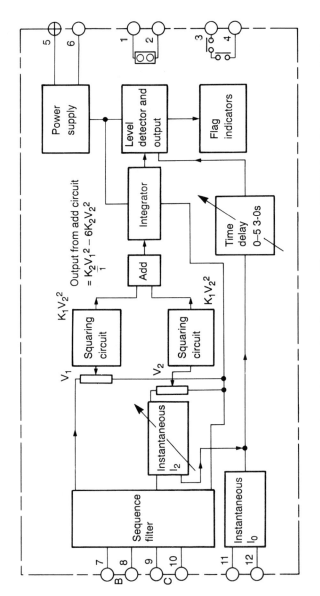

Figure 9.2 Schematic of a typical electronic thermal replica relay. (Courtesy of GEC Alsthom Measurements Ltd.)

considered too low to warrant the installation of specific protection, tripping was initiated by fire detection rather than upstream electrical protection devices.

Water from leaking pipes, or from driving rain or sea spray penetrating module walls, may enter switchgear enclosures, giving rise to explosive faults or serious fires. Arcing faults on the busbars of low-voltage switchgear will cause extensive damage to the switchboard and may lead to fire in an area of the installation likely to be close to accommodation modules or the process control room.

The offshore installation of a full bus zone protection scheme, with check and supervision relays, may only be advisable on the main switchboards of the largest platforms, owing to the weight of all the CTs required and the extra space taken up by such a scheme. Nevertheless, it is important that every part of the power system is adequately covered by the protection scheme, and the busbars of switchboards are no exception to this rule.

9.4.2 Overcurrent and directional overcurrent protection

The overcurrent fault protection relays on the primary circuit of the transformer will provide some protection to the secondary circuit. However, earth fault protection devices on the upstream side will not provide any secondary circuit protection as there is normally no zero-sequence path through the transformer.

If the busbars are sectionalized, the bus section switch may be fitted with overcurrent or directional overcurrent protection. Plain overcurrent relays operating on bus section switches cannot, of course, remove the faulted section from the supply, unless it is being fed via another section. However, if the switchboard is fed by plain feeders or directly by generators with overcurrent protection, such an arrangement can provide reasonable overcurrent protection whilst preserving the supplies on healthy bus sections (see Figure 9.3).

If the switchboard has three sections, directional overcurrent relays, arranged to detect fault currents flowing outwards, may be fitted to the bus section switches. Such an arrangement will preserve the two healthy sections provided the fault is not on the central section. As central section faults rely on upstream protection for clearance, this section should be kept as short and uncomplicated as possible for good reliability. This arrangement could be used for the connection of a third generator, with all outgoing feeders (except third-generator auxiliaries) fed from the end sections (see Figure 9.4).

Figure 9.3 Simple plain overcurrent busbar protection scheme

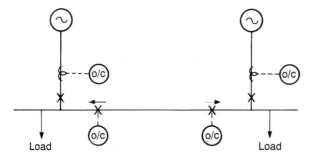

Figure 9.4 Simple directional overcurrent busbar protection scheme

9.4.3 Unrestricted earth fault protection

On main generator switchboards, either IDMT or definite time earth fault characteristics may be used. The use of definite time is often preferable, as fault current magnitudes are restricted by the generator neutral earthing resistors. Any relay used should be designed so as not to be sensitive to third-harmonic currents, if the generator winding configuration is such that significant third-harmonic currents may circulate.

Low-voltage switchboards are more likely to suffer arcing faults than medium-voltage switchboards. This is because, with increasing voltage, the reduction in fault current due to arc resistance becomes less pronounced as the arc voltage ceases to be a significant proportion of the total fault circuit driving voltage. Hence the importance of earth fault protection for low-voltage switchboards is stressed.

9.4.4 Frame earth protection

This type of system has been used successfully offshore, and consists of a frame earth relay and neutral check relay, arranged to monitor earth fault current flows into and out of the switchboard. A simple frame earth system is illustrated in Figure 9.5. Care must be taken, when installing the switchboard into the module, that it is and will remain insulated from the steel of the module construction. The generator earthing resistor and the switchboard frame should be earthed at the same point, to avoid series earth connections and to reduce the risk of the fault current being limited below the operating setting of the relay by a poor connection due to corrosion of the decking steel etc. This is less likely to be a problem than in an onshore substation using earth electrodes. However, earth connections should be maintained and checked for resistance value on a routine basis, particularly in order to prevent dangerous potentials appearing on the switchboard frame during fault conditions.

An earth or bonding bar interconnects the framework of each cubicle, but is only earthed via the conductor monitored by the frame earth relay. Where the circuit breakers are in trucks, the truck must also have a heavy current earth connection to the bonding bar in order to allow earth fault current flow from the circuit breaker.

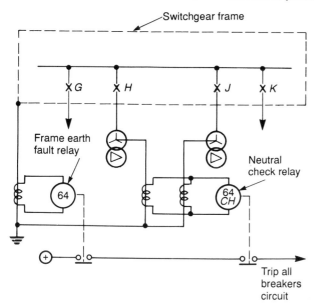

Figure 9.5 Busbar frame earth protection scheme. (Courtesy of GEC Alsthom Measurements Ltd.)

To prevent currents induced in cables from causing spurious tripping, the gland plates at the switchboard end should be insulated from the switchboard frame but earthed to a second separate earth bar which is directly earthed to the main earth point.

9.4.5 Differential protection

Kirchhoff's first law may be directly applied by monitoring the incoming and outgoing currents at all the switchboard external circuit connections, summing them and detecting the error, should an internal switchboard fault exist. The switchboard may be divided into a series of zones, and each zone boundary treated in the same way, as shown in Figure 9.6. The relay settings must be such that the transient flux produced by through fault currents will not cause tripping because of imbalance due to unequal burden or saturation of CTs.

By the use of high-impedance relays with series stabilizing resistors, the voltage across the relay during through faults, known as the stability voltage, may be kept below the relay operating voltage. The minimum value of the stabilizing resistor for stability may be calculated as follows:

$$V_S = \text{relay setting current} \times (R_{SR} + R_R)$$

where V_S is the stability voltage, R_{SR} is the resistance of the stabilizing resistor and R_R is the resistance of the relay coil. An example calculation is given in Section 9.8.

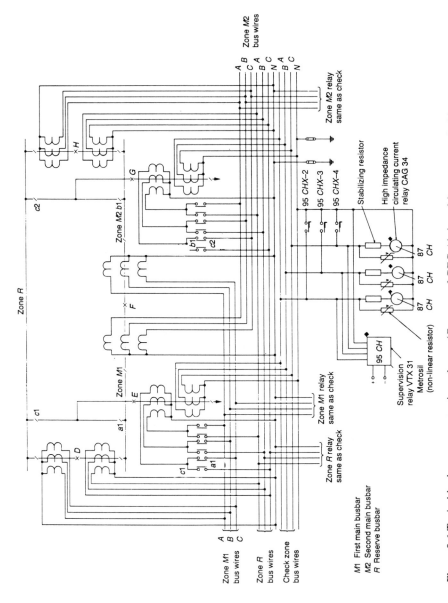

Figure 9.6 Typical busbar zone protection scheme. (Courtesy of GEC Alsthom Measurements Ltd.)

9.5   Feeder protection

9.5.1 Fuses

Low-voltage distribution systems in offshore installations are protected by fuses in the same way as their onshore counterparts. Although the *IEE Wiring Regulations* specifically excludes offshore installations from its scope (Part 1 paragraph 11-3(vii)), the document that it refers to (*IEE Recommendations for the Electrical and Electronic Equipment of Mobile and Fixed Offshore Installations*) is not sufficiently detailed for stand-alone use, and therefore both documents are used for guidance. The methods of calculation given in the *IEE Wiring Regulations* are normally adopted when designing lighting and small power distribution systems, particularly for accommodation modules.

The use and limitations of HRC fuses are discussed in Chapter 6.

9.5.2 Miniature circuit breakers (MCBs)

In most situations, miniature circuit breakers may be used as an alternative to fuses where the fault rating of the MCB is sufficiently high. However, because of the different shape of the MCB tripping characteristic compared with that of a fuse (see Figure 9.7), it is not advisable to mix fuses with MCBs in the same circuit if discrimination is required. The degree of discrimination between one MCB and another, and between MCBs and fuses, varies with the BS 3871 MCB type, but generally a better discrimination can be obtained using fuses.

Miniature circuit breakers with fault current ratings in excess of 16 kA are now available and are in common use offshore, where they save weight and reduce the quantity and type of spare fuses stocked. MCBs used offshore must provide positive indication of contact clearance, and must be padlockable when in the open position.

9.5.3 Overcurrent and earth fault protection

Most of the conventional forms of short cable feeder protection may be used offshore. The use of IDMT overcurrent and residually connected earth fault relays may give discrimination problems if a large proportion of the platform load is on the switchboard being fed. This is because the steady state fault current available from the platform generators will be of the order of three to four times the full load current; therefore, if fault clearance times are to be kept short, generator and feeder relay current settings will need to be close. Voltage controlled relays may be used on generator circuit breakers to overcome this problem, as discussed earlier for generator protection.

An alternative is to use definite time relays. However, on low-voltage systems it is less feasible to use definite time relays, since, if generator and main switchboard clearance times are to be kept reasonably short, there are likely to be discrimination problems with fuses and MCBs lower down in the system.

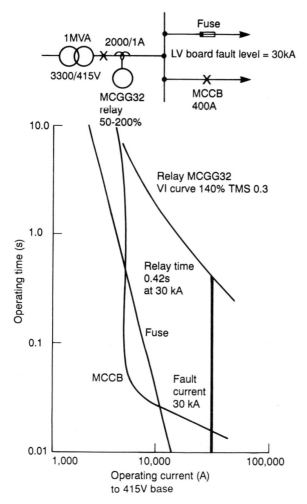

Figure 9.7 Miniature circuit breaker and fuse characteristic comparison. (Courtesy GEC Alsthom Measurements Ltd.)

As cables are more exposed to mechanical damage than switchboard busbars, it is advisable to protect cables which interconnect switchboards by some form of simple unit protection, rather than IDMT relays with intertripping. This has the advantage of faster operation and may also relieve any discrimination problems associated with the unrestricted method of protection.

9.6   Sizing of conductors

9.6.1 Load flow

When the electrical distribution system has been configured for optimum convenience, safety and reliability, the various busbars and cables need to be sized for the maximum continuous load in each system operating condition.

The first task is to ensure that the system 24 hour load profile and the load schedule are as up to date as possible, and that diversity factors and operating modes have been agreed by all parties and 'frozen'. If the system is simple, with few parallel paths, load flows may be manually calculated. In either steady state or transient conditions, the power system can be represented by a physical model, such as produced in a network analyser, or by a mathematical model using a digital computer. With the proliferation of desktop computers, the use of network analysers, even on small systems, is now rare.

On larger installations, with many parallel paths, computer load flow programs should be used in any case. Such programs are now available for use on desktop microcomputers at prices starting from a few hundred pounds.

Load flow calculations by nodal analysis have become firmly established. Such methods involve:

(a) the solution of a set of linear simultaneous equations which describe the system configuration;
(b) the application of restraints at each node to enable the required complex power and voltage conditions to be maintained.

The advantages in using nodal voltage analysis are that the number of equations is smaller than with the alternative mesh current analysis method, and the system may be described in terms of its node numbers and the impedances of the interconnecting branches. In nodal analysis, the node voltages V are related to the nodal injected currents I by the system admittance matrix Y. In matrix form:

$$[I] = [Y] [V]$$

The voltage V refers to a value between node and earth, and the current I is the injected nodal current. The total nodal injected power S is obtained from the product of voltage and current conjugate, as follows:

$$[S] = [V] [I]^*$$

By taking the current conjugate, the reactive power is given the same sign as active power for lagging current.

There are three basic types of nodal constraint:

(a) fixed complex voltage;
(b) fixed complex power;
(c) fixed voltage modulus with real power.

A type (a) constraint is given to the reference node, usually known as the 'slack' or (in the US) 'swing' bus. The type (b) constraint represents a load bus, and the type (c) a generation bus.

The formation of the nodal matrix, and methods available for digital iterative sequences of solution, are given in Bergen (1986).

9.6.2 Busbar sizing

Switchboard main busbars must be rated to carry the maximum continuous load which can flow in any healthy power system operating condition. Transient conditions giving rise to higher currents, such as those due to large motors starting or downstream faults, may be tolerated momentarily, provided that protection devices are incorporated which will ensure that the outgoing equipment is removed from the system before the busbars overheat.

The continuous current rating must be for the busbars as enclosed in their offshore environmental protection, with natural cooling only. This also applies to the switching and isolating devices in the switchboard.

9.6.3 Cable sizing

Cable sizing is generally carried out in accordance with the *IEE Wiring Regulations*, as would be the case onshore.

For motor cables, the basis of the calculation for voltage drop is

$$L = \frac{V_d \times 1000}{1.732 \times I \times (R\cos\theta + X\sin\theta)}$$

where L is a the cable length (m), V_d is the permitted steady state voltage drop (V), I is the motor full load current (A), R is the cable resistance (ohms/km), X is the cable reactance (ohms/km) and $\cos\theta$ is the motor power factor at full load. The permissible maximum steady state voltage drop is normally 2.5%, whilst the permissible voltage drop during starting is 10%.

A typical computer spreadsheet generated motor cable sizing chart is shown in Figure 7.14. De-rating factors for cables which pass through insulation and for bunching, application of protective devices etc. need to be considered in accordance with the IEE Wiring Regulations.

9.7 Worked example: fault calculation

The following calculations and information are not exhaustive, but are intended to give the reader sufficient knowledge to enable switchgear of adequate load and fault current rating to be specified. The subject may be studied in more detail by reading the relevant documents listed in Appendix 1. The nomenclature used is generally as given in the GEC Measurements *Protective Relay Application Guide* and the Electricity Council's *Power System Protection* (IEE).

When a short-circuit occurs in a distribution switchboard, the resulting fault current can be large enough to damage both the switchboard and associated cables owing to thermal and electromagnetic effects. The thermal effects will be proportional to the duration of the fault current to a large extent, and this time will depend on the characteristics of the nearest upstream automatic protective device which should operate to clear the fault.

Arcing faults due to water or dirt ingress are most unlikely in the switchboards of land based installations, but from experience they need to be catered for offshore. For switchboards operating with generators of 10 MW or more, it is usually not difficult to avoid the problem of long clearance times for resistive faults. However, with the smaller generators, clearance times of several seconds may be required because of the relatively low prospective fault currents available (see earlier section on busbar protection). With small emergency generators, pilot exciters are not normally provided and the supply for the main exciter is derived from the generator output. This arrangement is not recommended, as it allows the collapse of generator output current within milliseconds of the onset of a fault. With such small generators even subtransient fault currents are small, and it is unlikely that downstream protection relays set to operate for 'normal' generation will have operated before the output collapse. It is usual to provide a fault current maintenance unit as shown in Figure 9.8. This device is basically a compounding circuit which feeds a current proportional to output current back to the exciter field. When the output current reaches a threshold value well above normal load current, a relay operates, switching in the compounding circuit. Thus a high output current is maintained by this feedback arrangement until the operation of definite time overcurrent protection, set to prevent the generator thermal rating being exceeded.

9.7.1 Main generator protection

In this example, one generator is rated at 15.0 MW. The alternator has the following parameters: rating 17.7 MVA at a power factor of 0.85 and voltage of 6.6 kV; reactances $X_d'' = 0.195$ per unit (pu), $X_d' = 0.31$ pu, $X_d = 2.2$ pu, $X_q'' = 0.265$; full load current 1550 A; neutral earthing resistor 10 ohms. There are four more generators of a different type, each rated at 3.87 MVA and having parameters as follows: reactances $X_d'' = 0.15$ pu, $X_d' = 0.18$ pu, $X_d = 2.05$ pu, $X_q'' = 0.15$; full load current 340 A; neutral earthing resistor 40 ohms.

The generator reactance values (pu) may be tabulated on their own base as follows:

	X_d''	X_q''	X_o''	R_n
1 × 17.7 MVA unit	0.195	0.265	0.09	10 ohms
1 × 3.87 MVA unit	0.15	0.15	0.045	40 ohms

Referring to a common base of 17.7 MVA, 6.6 kV and considering the units in parallel:

	X_d''	X_q''	X_o''	R_n
1 × 3.87 MVA unit	0.686	0.686	0.2058	16.25
4 × 3.87 MVA units	0.1715	0.1715	0.0515	4.06
1 × 17.7 MVA unit	0.195	0.265	0.09	4.06

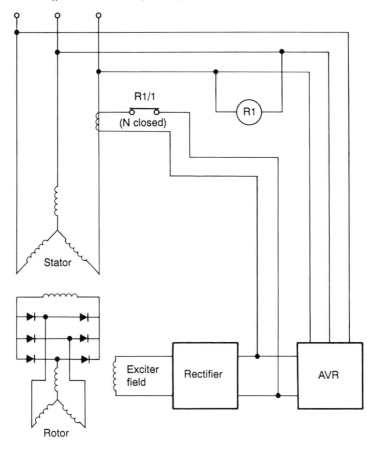

Figure 9.8 Schematic of small-generator fault current maintenance circuit

Summing the reactances:

$$X_d'' = \frac{1}{1/0.195 + 1/0.1715} = 0.0912 \text{ pu}$$

$$X_q'' = \frac{1}{1/0.265 + 1/0.1715} = 0.104 \text{ pu}$$

$$X_o'' = \frac{1}{1/0.09 + 1/0.0515} = 0.0327 \text{ pu}$$

$$R_n = \frac{1}{1/4.06 + 1/4.06} = 2.63 \text{ pu}$$

Hence:

$$X_d'' + X_q'' = 0.0912 + 0.104 = 0.1952 \text{ pu}$$
$$X_d'' + X_q'' + X_o'' = 0.912 + 0.104 + 0.0327 = 0.2279 \text{ pu}$$
$$|Z_t| = (X_d'' + X_q'' + X_o'')^2 + (3R_n)^2$$
$$Z_t = (0.2279)^2 + (3 \times 2.03)^2 = 6.09 \text{ pu}$$

Therefore the system fault currents (neglecting motor contribution) are:

three-phase fault current $= \dfrac{17.7}{3 \times 6.6. \times 0.0912} = 16.97\,\text{kA}$

phase-to-phase fault current $= \dfrac{17.7}{6.6 \times 0.1952} = 13.74\,\text{kA}$

Earth fault current $= \dfrac{3 \times 17.7}{6.6 \times 6.09} = 0.762\,\text{kA}$

This method of calculation provides low values which, although not suitable for providing fault currents for the selection of switchgear, are useful for relay setting, as neglecting motor fault contributions provides a safety margin. The extra fault current provided by motors should significantly reduce the operating times of all overcurrent and earth fault protection devices, provided that CT saturation is not a problem. A check should be made, however, that generator cable reactances do not reduce prospective fault currents by more than a negligible amount. This reduction is likely to be significant for the smaller generators operating at low voltages.

9.7.2 Switchboard protection

The prospective fault currents for medium-voltage switchboards can be obtained in a similar manner by summing generator and transformer reactances as follows. Take the base again as 17.7 MVA, and take as an example the generator switchboard connected to the LV switchboard by a 2 MVA transformer with 440 V secondary winding. Then the sequence reactances (pu) are as follows:

	X_1	X_2	X_0
At transformer rating	0.13	0.11	0.11
On the base MVA	1.1505	0.974	0.974

Summing generator and transformer reactances:

$X_1 = 0.0912 + 1.1505 = 1.2417$ pu
$X_2 = 0.104 + 0.974 = 1.078$ pu
$X_0 = 0.974 = 0.974$ pu

Hence:

$(X_1 + X_2) = 1.2417 + 1.078 = 2.3197$ pu
$(X_1 + X_2 + X_0) = 1.2417 + 1.078 + 0.974 = 3.294$ pu

Therefore the LV system fault currents are:

three-phase fault current $= \dfrac{17.7}{3 \times 0.44 \times 1.2417} = 18.7\,\text{kA}$

phase-to-phase fault current $= \dfrac{17.7}{0.44 \times 2.3197} = 17.34\,\text{kA}$

earth fault current $= \dfrac{3 \times 17.7}{0.44 \times 3.294} = 21.15\,\text{kA}$

9.7.3 Overcurrent protection

The full load current of the largest motor on the system is 600 A. The starting current of this motor is 1300 A. (Note that this is with an autotransformer starter on 80% tap, and gives a starting time with this particular motor of 10 seconds.) The maximum momentary fault current is $17.7/(0.195 \times 3 \times 6.6) = 7.94\,\text{kA}$ (RMS). The steady state fault current from manufacturer's test data is $3.54\,\text{kA}$ (after 0.25 seconds).

Let us select in this instance a CEE ITG 7231 relay for three-phase duty, using a range of $0.7–2.0\ I_n$. A CT ratio of 2000/1 A is selected, with 15VA5P10 CTs. The relay settings must coordinate with the following:

1. The total standing load of 200 A on the switchboard that the generator is supplying, plus the starting current taken by the 600 A motor. Therefore the load current is $200 + 1300 = 1500\,\text{A}$ for 10 seconds.
2. From simulation studies or voltage dip calculations, the maximum transient load for the maximum voltage dip of 80% (which must not be exceeded whilst this motor is being started) is obtained. This gives a load of 2010 A for an initially fully loaded generator.
3. The 600 A motor thermal overload relay. This is set to operate at $3.4\,\text{kA}$ after 10 seconds.

Therefore the minimum relay setting current required is $I_r = 2010/2000$ or approximately 1 A.

To coordinate with point 2 above, the operating time should be approximately $10.0 + 0.5 = 10.5$ seconds at $3.4\,\text{kA}$. The required plug setting multiple PSM is $3.4\times10^3/2010 = 1.69$ times setting. From the characteristic, at 1.69 times the relay setting and a time multiplier of 1.0, the relay operating time is 5 seconds. Hence, for a required operating time of 10.5 seconds, the required time multiplier T_m is $10.5/5 = 2.1$.

9.7.4 Digital methods of fault calculation

In digital fault calculation, an admittance matrix is formed which is extended to include the source admittances, and the matrix is then reduced to a single impedance connected between the neutral (zero node) and the point of fault. A simple fault calculation program is given in Appendix C, using PC compatible GWBASIC.

9.7.5 Digital simulation of system disturbances

When designing large offshore power systems, it is vital that the response of the system to the starting of large motors, and the application of phase-to-phase or earth faults, have been analysed to a reasonable degree of certainty, so that the risk of dangerous operating conditions and system instability can be avoided as much as possible.

The digital computer programs used in such analysis have the following elements:

1. The program will obtain its basic system data, including models of all generators, governors, AVRs etc. and network configuration, from

data stored in various input files. Models for such devices as AVRs, prime movers etc. are in the form of differential equations.

2. Having obtained the necessary input data and initialized the system, the program uses a step-by-step implicit trapezoidal method of simultaneously solving the large number of differential equations involved. Iterative methods are used at each step to obtain the desired accuracy (usually 99.95%).

3. After a defined simulated period, the program stops and creates an output file containing the values of all variables calculated at each step.

4. For faster and more user-friendly interpretation of this output data, graphics subroutines are used to plot the variables against time.

Suitable program suites are available commercially from the organizations listed in Appendix D.

9.8 Worked example: relay setting of typical MV platform scheme

To avoid repetition, only the coordination of the system operating in mode 1 will be considered. The relay and fuse coordination formulae have been obtained from the GEC Measurements *Protective Relay Application Guide*.

9.8.1 Data requirements

9.8.1.1 System data
Before beginning work on such a scheme, ensure that you have at least the following information at hand:

(a) load flow study results for each operating configuration to be considered;

(b) short-circuit study results for each operating configuration to be considered, including three-phase-faults and phase-to-earth faults;

(c) single-line diagram of the system, identifying circuit breakers and their protective devices, fuses and any other means of fault interruption (see Figure 9.9);

(d) fault decrement curves for any generator sets, and fault levels for any incoming subsea cable feeders.

9.8.1.2 Base values
The following base values will be assumed:

Base MVA $= 10\,\mathrm{MVA}$
Base impedance at $440\,\mathrm{V} = 0.44^2/10 = 0.0194$ ohms
Base impedance at $6.6\,\mathrm{kV} = 6.6^2/10 = 4.356$ ohms
Base current at $440\,\mathrm{V} = 10\,\mathrm{MVA}/[(\sqrt{3}) \times 0.44\,\mathrm{kV}] = 13.12\,\mathrm{kA}$
Base current at $6.6\,\mathrm{V} = 10\,\mathrm{MVA}/[(\sqrt{3}) \times 6.6\,\mathrm{kV}] = 0.875\,\mathrm{kA}$
Z_{pu} for base 1 $= Z_{\mathrm{pu}}$ for base 2 $\times \mathrm{MVA}_{\mathrm{base}_1}/\mathrm{MVA}_{\mathrm{base}_2}$

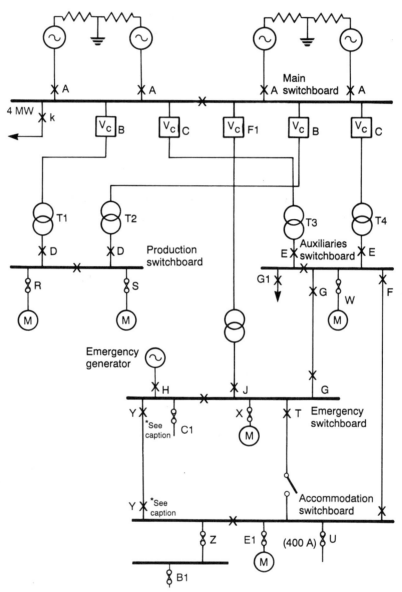

Figure 9.9 System diagram. Setting details given for relays T, H, G, F, Z only. Relay E is earth fault only; overcurrent protection is provided by relay C. Circuit breakers either side of T3 are intertripped. Relays marked * are not considered, since this feeder is not used in operating mode 1

9.8.1.3 Synchronous machine parameters

Table 9.1

Description	Per unit value to machine base		Per unit value to 10 MVA base	
	Production generator	Emergency generator	Production generator	Emergency generator
Rated MVA	4.729	1.313		
Rated voltage (kV)	6.6	0.44		
Subtransient reactance X''_d	0.181	0.233	0.383	1.775
Transient reactance X'_d	0.252	0.46	0.533	3.503
Synchronous reactance	1.95	2.45	4.123	18.66
Potier reactance	0.166	0.22	0.351	1.676
Saturation factor	1.45	1.562	1.45	1.562
Negative-sequence reactance	0.255	0.27	0.539	2.056
Zero-sequence reactance	0.103	0.14	0.218	1.066

9.8.1.4 Operating conditions

Mode 1 Maximum generation: four production generators, maximum standing load and all motors running. Main switchboard bus section closed, other switchboard bus sections open.

Mode 2 Minimum generation: one production generator running with no motors operating.

Mode 3 Emergency generation only.

The low-voltage switchboards are provided with two feeders. The normal configuration is with low-voltage switchboard bus section switches open, with each side fed by a separate feeder.

From Figure 9.9 an overcurrent discrimination flow chart can be constructed, as in Figure 9.10. All the devices including fuses should be identified, and the flow is downstream from left to right. The coordination exercise can now begin with the relays furthest downstream. It should not be necessary to start with the fuses U and Z as these obviously have fixed characteristics. However, it may still be necessary to change the fuses for ones with more suitable characteristics if grading problems are experienced.

9.8.2 Overcurrent relay setting

9.8.2.1 Relay F: range 10–200%, CT ratio 1500/1

From the load flow study:

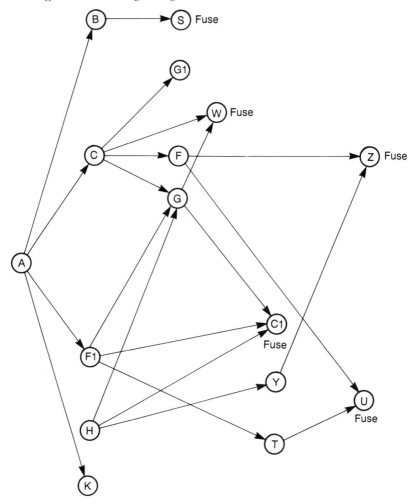

Figure 9.10 Overcurrent discrimination flow chart

standing load = 50 kW + j20 kVAr ≈ 71 A
motor load (2 motors) = 96 kW + j40 kVAr
motor load (1 motor) = 96/2 + j40/2 = 52 kVA = 68 A
motor starting current ≈ 6 × 68A = 408 A

Therefore the maximum load including motor starting, assuming that the currents are in phase, is

71 + 68 + 408 = 547 A

The largest fuse on the accommodation switchboard is 400 A (fuse U). The relay must discriminate with fuse U and permit the maximum load current of 547 A to flow.

For discrimination with the fuse, the primary current setting should be approximately 3 times the fuse rating, i.e. $3 \times 400 = 1200$ A. Therefore the nearest relay setting is $0.75 \times 1500 = 1125$ A (0.75 PSM).

The maximum fault level for fuse grading is 56.009 kA (from computer SC study). Therefore from the manufacturer's fuse characteristics, the fuse operating time t at 56 kA is 0.01 seconds. From the GEC *Guide*, the minimum relay operating time is given by

$$t + 0.4t + 0.15 = 0.164 \text{ seconds}$$

If the fault current from computer studies is 51.44 kA, then the fault current as a multiple of the plug setting is $51\,440/1125 = 46$. The relay operating time at 46 times the plug setting with a time multiplier setting (TMS) of 1 is 0.255 seconds, from the relay characteristic. Therefore the required TMS is $0.164/0.255 = 0.64 \approx 0.7$.

9.8.2.2 Relay T: range 10–200%, CT ratio 1500/1
As the maximum loading of the accommodation switchboard is the same, the exercise is similar to that for relay F but using the fault currents that the computer has calculated for feeder T.

With the given fault current of 44 280 A, the equivalent relay PSM is $44\,280/1125 = 39.4$. Therefore from the relay characteristic, using a PSM of 39.4 and a TMS of 1, the operating time is 0.255 seconds. The required TMS is $0.164/0.255 = 0.64 \approx 0.7$.

9.8.2.3 Relay G: range 10–200%, CT ratio 2000/1
In mode 1, the utilities switchboard would be feeding the emergency switchboard, as the emergency generator would not be running. From the load flow data for the emergency switchboard:

load $= 455$ kW $+$ j193 kVAr ≈ 649 A at 0.92 power factor
largest motor load ≈ 84 A at 0.8 power factor
motor starting current ≈ 504 A at 0.2 power factor

Therefore the maximum load during the starting of this motor is

$$\text{load} - \text{ motor FLC } + \text{ motor start}$$
$$(597 + \text{j}253) - (67.2 + \text{j}50.4) + (100.8 + \text{j}494)$$
$$= 939.6 \text{ A at } 0.67 \text{ power factor} \approx 1000 \text{ A}$$

Relay G must permit the maximum load without tripping and must also coordinate with the largest outgoing fuse on the emergency switchboard, i.e. fuse C1 (160 A) must operate before relay G. Coordination with relay T is not considered necessary, since the accommodation switchboard would be supplied through circuit F. Relay Y does not need to coordinate with relay G, since feeder Y is only used in mode 3 operation (emergency generator only). However, the setting of relay G must also allow for back feeding of the utilities switchboard via the emergency switchboard and therefore coordination with fuse W, the largest utilities switchboard fuse.

The maximum fault current through feeder G is 56 000 A (from fault calculations). The operating time t of fuse C1 at 56 kA is 0.01 seconds, from the fuse characteristic. The setting of relay G for coordination with fuse C1 is 3 times the fuse rating, i.e. 3×160 A $= 480$ A, but the maximum load is 1000 A.

For back feeding of the utilities switchboard, relay G must coordinate with fuse W, which gives 3×355 A $= 1065$ A. From the load flow results, the maximum current flow through feeder G in the direction of the utilities switchboard is 1032 A at 0.8 power factor.

The starting current of the largest motor is $261 + j1281$. Therefore the maximum load is

$$(825.6 + j784.8) - (174.4 + j130.8) + (261 + j1281) = 912.2 + j1935 = 2139 \text{ at } 0.43 \text{ power factor}$$

Therefore use the nearest setting of 2000 A. This is also the rating of the transformer which feeds the emergency switchboard.

From the SC calculations, the maximum SC current at fuse W is 44.3 kA. From the fuse characteristics, the fuse operating time t is 0.01 seconds. Therefore relay G operating time is

$$t + 0.4t + 0.15 = 0.164 \text{ seconds}$$

The fault current as a multiple of the relay plug setting is $44\,300/2000 = 22$. Therefore from the relay characteristic the operating time of relay G at a TMS of 1 is 0.3 seconds. The required TMS is $0.164/0.3 = 0.54 \approx 0.6$ (nearest upward setting; using a setting of 0.5 would reduce the grading margin).

9.8.2.4 Coordination
The same methods may be adopted to obtain settings for the other overcurrent relays. The resulting coordination chart is shown in Figure 9.11.

9.8.3 Earth fault relay setting
Figure 9.12 shows the earth fault discrimination flow chart for this system.

9.8.3.1 CT saturation
It is necessary to check for CT saturation where it is considered to be a likely condition. The saturated CT current may be obtained as follows:

$$I_{\text{max sat}} = \frac{\text{CT rated VA} \times \text{accuracy limit factor} \times \text{CT ratio}}{\text{secondary burden at rated current}}$$

If the maximum fault current available is less than this, coordination should not be lost owing to CT saturation.

9.8.3.2 Relay Z: range 10–20%, CT ratio 300/1. Fuse Z: 200 A
Relay Z and fuse Z must coordinate with the largest fuse in the next downstream switchboard, which is 125 A (B1 on the coordination chart). From the fuse catalogue: I^2t (pre-arcing) of fuse Z is 2×10^5 A^2 s; I^2t (total operating) for fuse B1 is 9×10^4 A^2 s. Therefore as $I^2t_Z \gg I^2t_{B1}$, coordination is assured.

Relay Z operates at lower fault currents while fuse Z operates at higher fault currents. Relay Z must also discriminate with fuse B1, however.

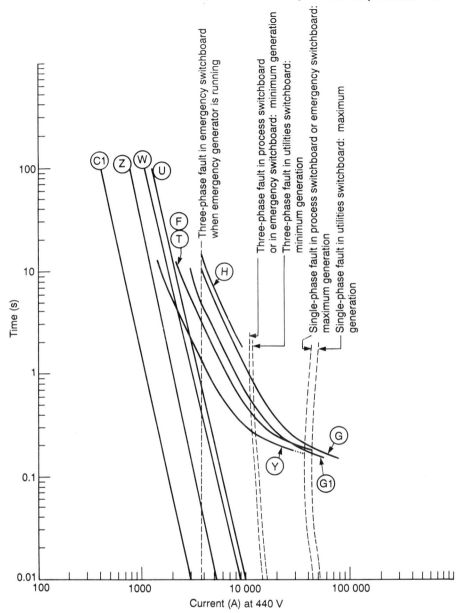

Figure 9.11 Overcurrent protection coordination chart – low voltage system

Therefore set relay Z at approximately 3 times the rating of fuse B1, i.e. 3 × 125 = 375 A; then the minimum relay setting is 1.5 × 300 = 450 A.

Fuse B1 operates at $t = 0.01$ s at 2400 A, from the fuse characteristic, with fault current obtained from computer results. From the GEC *Guide*, the relay Z operating time is

$$t + 0.45t + 0.15 = 0.164 \text{ seconds}$$

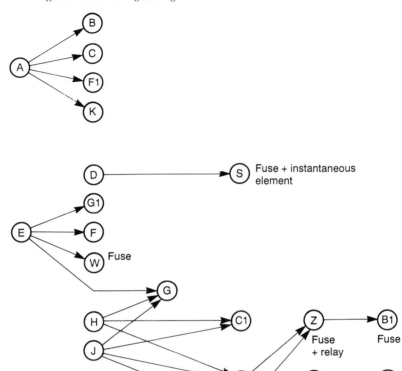

Figure 9.12 Earth fault discrimination flow chart

The PSM is 2400/450 = 5.33. Therefore the relay operating time at a PSM of 5.33 and a TMS of 1 is 1.8 seconds, from the relay characteristic. The required TMS is 0.164/1.8 = 0.09 ≈ 0.1.

In this case, a check was made against the relay characteristic. It was found that the curve for TMS = 0.1 gives an operating time of only 0.14 seconds at a PSM of 5.33. As this is too fast, the time setting was adjusted to 0.2.

Relays U, T and Y may be set in a similar manner.

9.8.3.3 Relay F: range 10–200%, CT ratio 1500/1; 10 VA 5P10

This relay must coordinate with earth fault relays Z and U and with the largest outgoing fuse in the accommodation switchboard, fuse U, 400 A. (This assumes that coordination with fuse U assures coordination with relay.)

Set relay F at 3 times fuse U rating, i.e. 3 × 400 A = 1200 A. Then the nearest current setting is 0.8 × 1500 = 1200 A. From the SC studies, the maximum earth fault current is 55.458 kA. At this current, fuses Z and U operate in 0.01 seconds, from the fuse characteristics. From the GEC *Guide*, the relay F operating time is

$t + 0.4t + 0.15 = 0.164$ seconds

From the SC studies, the average earth fault current at 0.164 seconds is 51.44 kA. As a PSM, this is $51\,440/1200 = 42.9$. From the relay character-istics, the relay operating time at a PSM of 42.9 and a TMS of 1 is 0.25 seconds. Therefore the required TMS is $0.164/0.25 = 0.66 \approx 0.7$.

It is necessary to carry out a relay saturation check. The relay impedance at the 0.8 setting is given by $(\text{relay VA})/(0.8)^2 = 0.39$ ohms. Allowing 1 ohm for lead resistance gives 1.39 ohms. Thus the secondary burden is $1.39 \times (1\text{ A}) = 1.39$ VA. Therefore

$$I_{\text{max unsat}} = 10\text{ VA} \times (10/1.39) \times (1500/1) = 107.9\text{ kA}$$

which is above the maximum fault current. Note that if a saturation current significantly lower than the fault current is obtained, it is advisable to select a different CT. The burdens associated with electronic relays are much lower than those used above for induction disc types, and because of this are less likely to give rise to saturation problems.

9.8.3.4 Coordination
The same method may be used for the setting of the remaining relays. The resulting coordination chart is shown in Figure 9.13.

9.9 Worked example: setting voltage controlled overcurrent relays

Some overcurrent relays provide dual inverse characteristics. The less sensitive characteristic gives a longer time interval for clearance of downstream faults, and the more sensitive characteristic removes close-up faults more quickly. Faults close to the generator have less impedance and hence produce greater voltage dips. At a set level of voltage dip, the relay will switch from one characteristic to the other.

In this case we will set the voltage to discriminate between a fault upstream (giving rise to a large voltage dip) and a fault downstream of a transformer (producing a smaller voltage dip). The inset on Figure 9.14 shows the equivalent diagram. The maximum voltage dip occurs when the transformer reactance X_t is a minimum compared with the generator reactance X_g. In this case we will assume this is when one generator is operating and two transformers are paralleled. The voltage at the generator (high-voltage) terminals is given by

$$V = \frac{X_t}{(X_t + X_g)} - E$$

where E is the generated voltage.

The generator reactances from say 0 to 2 seconds can be calculated from the generator decrement curves or from values given by the manufacturer. In this case the generators are identical, each with reactances of $15.1\%_{t=0}$ and $33\%_{t=1}$, and each transformer has a reactance of 20.12% at the same MVA base. Therefore with one transformer:

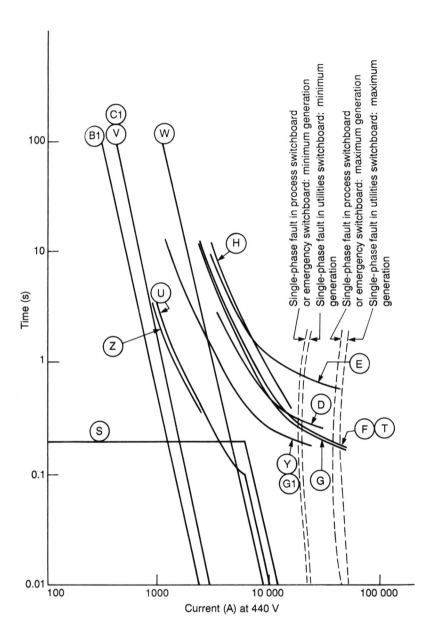

Figure 9.13 Earth fault protection coordination chart – low voltage system

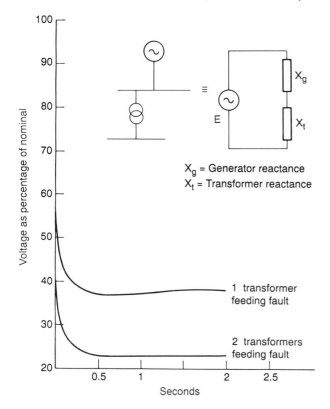

Figure 9.14 Graph of voltage downstream of transformers during system fault: Example 9.9.

$$V_{t=0} = 20.12/(20.12 + 15.1) = 57\%$$
$$V_{t=1} = 20.12/(20.12 + 33) \quad = 38\%$$

With two transformers in parallel:

$$V_{t=0} = 10.06/(10.06 + 15.1) = 40\%$$
$$V_{t=1} = 10.06/(10.06 + 33) \quad = 23.4\%$$

Therefore a setting of 20% should prevent the relay from switching to the more sensitive characteristic on a low-voltage fault.

Two further points may be made. First, this type of relay is known as a voltage controlled overcurrent relay, and should not be confused with the voltage restrained type where the sensitivity is continuously variable over a voltage range. The voltage restrained type is not recommended since sensitivity increases too quickly with dipping voltage so that discrimination with downstream fuses is lost. Secondly, some American relays are also known as voltage controlled, but with these the inverse characteristic is inhibited at low levels of voltage dip, leaving only a definite time delay element to clear the fault.

Chapter 10

Offshore lighting

If those readers unfamiliar with offshore conditions imagine being woken up in the top bunk of a pitch black cabin by a strange alarm sounding, possibly accompanied by the sound and shock of explosions, they will appreciate that every aid to orientation and escape is vital. (Surprisingly, North Sea operators have not yet standardized on alarm sounds.)

Particularly since the Alexander Kielland disaster some years ago, when an accommodation semisubmersible capsized at night with heavy loss of life, offshore engineers have been aware of the importance of good escape lighting. Visual information from our surroundings is of vital importance in any environment if comfort and safety are to be maintained. In the offshore environment, bad lighting can very easily lead to accidents and injury and should be considered as an essential topic of any design safety audit.

10.1 Lighting calculations

10.1.1 Point sources

In the visible wavelength range, radiant flux or electromagnetic radiation is considered to have a *luminous flux* associated with it. This luminous flux is a measure of human visual response, and a point source of light emitting a uniform intensity of 1 candela in all directions emits a total flux of 4π lumen (lm).

The illumination effect of a point source of light is shown in Figure 10.1. A point source S, emitting luminous flux in all directions, illuminates a plane surface P. The flux $d\phi$ intercepted by an element of area dA on P is the flux emitted within the solid angle $d\omega$ subtended at the source by the element dA, assuming no absorption of light takes place in the space between the source and the surface. The term $d\phi/d\omega$ is called the *luminous intensity* I of the source of the direction being considered. The accepted unit for luminous intensity I is the candela (cd) or lumen per steradian; the imperial unit used prior to this was the candlepower.

The *illuminance*, denoted by E, is the luminous flux falling on a surface. The unit of illuminance is the lumen per square metre, such that

$$E = \frac{I\cos\theta}{r^2} \qquad (10.1)$$

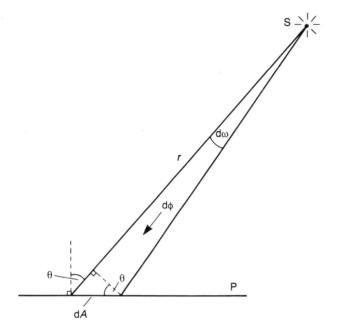

Figure 10.1 Illumination effect of a point source of light

The inverse square law is only strictly applicable to point sources of light, but can be applied for practical purposes to large diffusing sources providing that r is more than five times the largest dimension of the source.

A typical point source calculation sheet is shown in Figure 10.2.

10.1.2 Linear sources

Large fluorescent lamps are obviously not point sources, and any calculation must take this into account. In Figure 10.3 a linear source of length l is represented by AB. The illuminance is required on P in the plane CD which is parallel to the source. APEB is a plane which passes through the axis of the source, is at an angle θ to the vertical, and makes an angle ϕ with the normal PN to the plane CD.

Consider an element δx at S, distant x from A. ST lies in the axial plane and is normal to the axis of the source. Let $I(\alpha,\theta)$ be the intensity of the source AB in the direction parallel to SP. The illuminance δE at P on the plane CD due to element δx of the source is then given by

$$\delta E = \frac{\delta x I(\alpha,\theta)}{l} \frac{\cos \alpha \cos \phi}{(x/\sin\alpha)^2} \tag{10.2}$$

Since

$$AP = \frac{h}{\cos \theta} = \frac{x}{\tan \alpha}$$

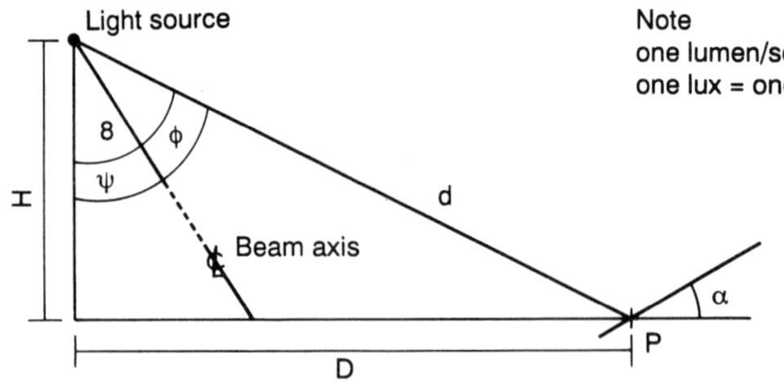

Note
one lumen/sqft = 10.76 lux
one lux = one lumen/sq m

Calculation
Location_______________ drawing no_________rev___
Required illumination at point P____________lux________

	A	B	C	
Fixing reference no.______________				
Mounting height above point P__________(H)				Metres/ feet
Distance from source______________(D)				Metres/ feet
Angle of inclined plane_____________(α)				Degrees
Beam axis angle to vertical___________(ψ) (angled fitting only)				Degrees
Luminous intensity from polar__________(I_o) Curve at angle 8 (vertical fitting) or angle ϕ (angled fitting)				Degrees

A
Fitting reference no. $\tan 8 = \frac{D}{H}$ = ______ ∴ 8 = [ ] Degrees
$d = D^2 + H^2$ = _____ + _____ ∴ d^2 = [ ]
Illumination at point = $\frac{I_o \cos(8-\alpha)}{d^2}$ = ______ = [ ] lux/lum sq ft

B
Fitting reference no. $\tan 8 = \frac{D}{H}$ = ______ ∴ 8 = [ ] Degrees
$d = D^2 + H^2$ = _____ + _____ ∴ d^2 = [ ]
Illumination at point = $\frac{I_o \cos(8-\alpha)}{d^2}$ = ______ = [ ] lux/lum sq ft

C
Fitting reference no. $\tan 8 = \frac{D}{H}$ = ______ ∴ 8 = [ ] Degrees
$d = D^2 + H^2$ = _____ + _____ ∴ d^2 = [ ]
Illumination at point = $\frac{I_o \cos(8-\alpha)}{d^2}$ = ______ = [ ] lux/lum sq ft

Total illumination at point ________________ = [ ] lux

Lighting fitting data
Manufacturer & catalogue reference no..______
____
____
____
____
____

Description ______
____
____
____
____
____
____
____
____
____

Sketch

Figure 10.2 Typical point source calculation sheet

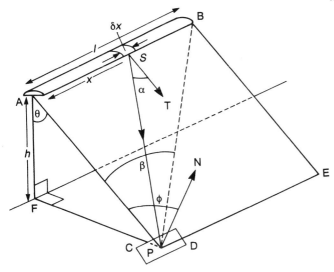

Figure 10.3 Illumination effect of a linear source of light

then

$$x = h \tan \alpha \sec \theta$$
$$\delta x = h \sec^2 \alpha \sec \theta \, \delta \alpha$$

Substitution of these for x and δx in Equation (10.2) gives

$$\delta E = \frac{I(\alpha,\theta)}{lh} \cos \theta \cos \phi \cos \alpha \, \delta \alpha$$

For the whole length of the source,

$$E = \frac{I(0,\theta)}{lh} \cos \theta \cos \phi \int_0^\beta \frac{I(\alpha,\theta)}{I(0,\theta)} \cos \alpha \, d\alpha \qquad (10.3)$$

The integral, known as the *parallel plane aspect factor*, is a function of the shape of the intensity distribution in the axial plane inclined at θ to the horizontal and the angle β (the aspect angle) subtended by the luminaire at P, which must be opposite one end of the luminaire, as in Figure 10.3.

The shape of the intensity distribution is similar in any axial plane for most fluorescent luminaires. For a given luminaire the aspect factor is independent of θ, varying only with β, so that it may be denoted by $AF(\beta)$.

Most diffuser cross-section shapes can be expressed mathematically, so that $AF(\beta)$ can be found by integration. For a uniform diffuser, for which $I(\alpha)$ equals $I(0) \cos \alpha$,

$$AF(\beta) = \int_0^\beta \cos^2 \alpha \, d\alpha$$

$$= \beta + \sin \beta \cos \beta \qquad\qquad (10.4)$$

A numerical method of integration may be used, such as dividing the axial curve into a number of equiangular zones.

To calculate the illuminance produced at a point opposite the end of the luminaire on a plane parallel to the luminaire (CD in Figure 10.3), the aspect angle β is determined and the value of $AF(\beta)$ is entered in Equation (10.3) to give

$$E = \frac{I(0,\theta)}{lh} \cos \theta \cos \phi \, AF(\beta) \qquad\qquad (10.5)$$

Should the point concerned not be opposite the end of the luminaire, the principle of superposition can be applied. If it is opposite a point on the luminaire then the illuminance due to the left and right hand parts of the luminaire are added. If it is beyond the luminaire then the illuminance due to a luminaire of extended length is reduced by the illuminance due to the imaginary extension.

Figure 10.4 shows a typical calculation sheet for linear source lighting.

The same method may be used for calculating the illuminance on a plane perpendicular to the axis of the luminaire (AFP in Figure 10.3). The derivation followed above applies, with $\sin \alpha$ replacing $\cos \alpha \cos \phi$ in Equation (10.2), resulting in a final expression for the illuminance at P of

$$E = \frac{I(0,\theta) \cos \theta}{lh} \int_0^\beta \frac{I(\alpha,\theta)}{I(0,\theta)} \sin \alpha \, d\alpha \qquad\qquad (10.6)$$

The integral expression is now the perpendicular plane aspect factor $af(\beta)$. Again this is substantially independent of θ in most practical situations, resulting in

$$E = \frac{I(0,\theta)}{lh} \cos \theta \, af(\beta) \qquad\qquad (10.7)$$

which determines the illuminance on a plane perpendicular to a linear luminaire.

10.1.3 Utilization factors for interiors and average illuminance

The utilization factor $UF(S)$ for a surface S is the ratio of the total flux received by S to the total lamp flux of the installation. Utilization factors are used to calculate the number of luminaires needed to provide a given illuminance on a surface. Utilization factors vary according to the light distribution of the luminaire, the geometry of the room, the layout of the luminaires, and the reflectance of the reflecting surfaces. The average horizontal illuminance $E_h(S)$ produced by a lighting installation, or the number of luminaires required to achieve a specific average illuminance,

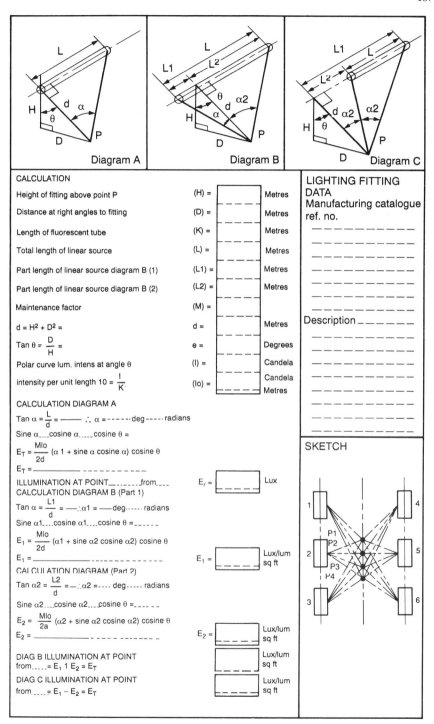

Figure 10.4 Typical calculation sheet for linear sources

can be calculated by means of utilization factors using the lumen method formula as follows:

$$E(S) = \frac{F \times n \times N \times MF \times UF(S)}{A_h(S)}$$

where F is the initial bare lamp flux, n is the number of lamps per luminaire, N is the number of luminaires, MF is the maintenance factor, associated with the deterioration due to dirt, dust and lamp ageing, $A_h(S)$ is the area of the horizontal reference surface S, and $UF(S)$ is the utilization factor for the reference surface S.

Although utilization factors may be calculated by the lighting designer (see Bibliography), most manufacturers publish utilization factors for standard conditions of use and for three main room surfaces. The first of these surfaces, the C surface, is an imaginary horizontal plane at the level of the luminaires, having a reflectance equal to that of the ceiling cavity. The second surface, the F surface, is a horizontal plane at normal working height, which is usually assumed to be 0.85 m above the floor. The third surface, the W surface, consists of all the walls between the C and F planes. Typical utilization factors are shown in Table 10.1.

The *room index* is a measure of the angular size of a room, and is the ratio of the sum of the areas of the F and C surfaces to the area of the W surface, each rectangular area of room being treated separately. For rectangular rooms, the room index is given by

$$RI = \frac{LW}{(L + W)H}$$

where L is the room length, W is the room width, and H is the height of the luminaire plane above the horizontal reference plane.

The *effective reflectances* (ratios of reflected flux to incident flux) are needed for the wall surface, the ceiling cavity and the floor cavity. The wall surface will consist of a series of areas A_1 to A_n of different reflectances R_1 to R_n respectively. The effective reflectance of a composite surface is the area-weighted average R_a, given by

$$R_a = \sum_{k=1}^{n} R_k A_k \bigg/ \sum_{k=1}^{n} A_k$$

The CIBSE Technical Memorandum TM5 provides a table of effective reflectances R_e, but an approximate value can be obtained from

$$R_e = \frac{CI \times R_a}{CI \times 2(1 - R_a)}$$

where R_a is the area-weighted average reflectance of the surfaces of the cavity and CI is the cavity index, defined as

$$CI = \frac{LW}{(L + W)d}$$

where d is the depth of the cavity.

Table 10.1 Utilization factors and initial glare indices

Utilization factors *UF*[F] Nominal *SHR* = 2.00

Room reflectances			Room index								
C	W	F	0.75	1.00	1.25	1.50	2.00	2.50	3.00	4.00	5.00
0.70	0.50	0.20	0.44	0.55	0.59	0.63	0.68	0.72	0.74	0.77	0.80
	0.30		0.38	0.49	0.54	0.58	0.64	0.68	0.70	0.74	0.77
	0.10		0.34	0.45	0.50	0.54	0.60	0.64	0.67	0.71	0.74
0.50	0.50	0.20	0.40	0.50	0.54	0.57	0.61	0.64	0.66	0.69	0.71
	0.30		0.35	0.46	0.49	0.53	0.58	0.61	0.63	0.66	0.69
	0.10		0.31	0.42	0.46	0.49	0.55	0.58	0.61	0.64	0.67
0.30	0.50	0.20	0.36	0.45	0.48	0.51	0.55	0.57	0.59	0.61	0.63
	0.30		0.32	0.42	0.45	0.48	0.52	0.55	0.57	0.59	0.61
	0.10		0.29	0.39	0.42	0.45	0.50	0.53	0.55	0.58	0.60
0.00	0.00	0.00	0.25	0.34	0.36	0.39	0.42	0.44	0.46	0.48	0.49
BZ class			4	3	3	3	4	4	4	4	4

Glare indices

Ceiling reflectance	0.70	0.70	0.50	0.50	0.30	0.70	0.70	0.50	0.50	0.30
Wall reflectance	0.50	0.30	0.50	0.30	0.30	0.50	0.30	0.50	0.30	0.30
Floor reflectance	0.14	0.14	0.14	0.14	0.14	0.14	0.14	0.14	0.14	0.14

Room dimension		Viewed crosswise					Viewed endwise				
X	Y										
2H	2H	7.0	8.4	8.0	9.5	10.8	6.8	8.2	7.8	9.2	10.5
	3H	8.9	10.2	10.0	11.3	12.6	8.6	9.8	9.6	10.9	12.2
	4H	9.9	11.1	10.9	12.2	13.5	9.4	10.6	10.4	11.7	13.0
	6H	11.0	12.1	12.0	13.2	14.5	10.3	11.4	11.3	12.5	13.8
	8H	11.6	12.6	12.6	13.7	15.1	10.7	11.8	11.7	12.9	14.2
	12H	12.2	13.2	13.2	14.3	15.7	11.1	12.1	12.1	13.2	14.6
4H	2H	7.7	8.9	8.7	10.0	11.3	7.5	8.7	8.5	9.8	11.1
	3H	10.0	11.0	11.0	12.1	13.5	9.6	10.6	10.7	11.7	13.1
	4H	11.2	12.1	12.2	13.2	14.6	10.6	11.6	11.7	12.7	14.1
	6H	12.5	13.4	13.6	14.5	15.9	11.8	12.6	12.8	13.7	15.1
	8H	13.3	14.0	14.4	15.2	16.6	12.3	13.1	13.4	14.2	15.7
	12H	14.0	14.8	15.1	15.9	17.3	12.8	13.5	13.9	14.7	16.1
8H	4H	11.8	12.6	12.9	13.7	15.2	11.4	12.2	12.5	13.3	14.7
	6H	13.5	14.2	14.6	15.3	16.8	12.8	13.5	13.9	14.6	16.1
	8H	14.4	15.0	15.5	16.1	17.6	13.5	14.1	14.6	15.2	16.7
	12H	15.4	16.0	16.6	17.1	18.6	14.2	14.8	15.4	15.9	17.4
12H	4H	12.0	12.7	13.1	13.8	15.3	11.6	12.3	12.7	13.5	14.9
	6H	13.7	14.3	14.9	15.5	17.0	13.1	13.7	14.3	14.9	16.4
	8H	14.8	15.3	16.0	16.5	18.0	14.0	14.5	15.1	15.7	17.2
	12H	15.7	16.2	16.8	17.3	18.8	14.6	15.0	15.7	16.2	17.7

Conversion terms:

Luminaire length (mm)	1500	1800
Wattage (W)	1 × 65	1 × 75
Conversion factor *UF*	1.00	1.00
Glare indices conversion	0.63	0.00

The ratio of the spacing between luminaires and their height above the horizontal reference plane, the spacing-to-height ratio (*SHR*), affects the uniformity of illuminance on that plane. The nominal and maximum values of *SHR* should be quoted in the *UF* tables.

10.2   Calculation procedure

10.2.1 Average illuminance

The following sequence of calculations is recommended in order to calculate the number of luminaires required to achieve a given average reflectance:

1.   Calculate the room index *RI*, the floor cavity index *CI*(F), and the ceiling cavity index *CI*(C).
2.   Calculate the effective reflectances of the ceiling cavity, the walls and the floor cavity. These must include the effects of furniture and/or equipment.
3.   Determine the *UF*(F) value from the manufacturer's tables, using the room index and the effective reflectances calculated.
4.   Calculate the maintenance factor *MF* by multiplying together the various factors obtained from the manufacturer's curves which reduce effective luminaire output.
5.   Use the lumen method formula to obtain the number of luminaires required.
6.   Arrange the lamps in a suitable configuration for the room. If the furniture or equipment in the room requires variations in spacing, refer to the method given in CIBSE TM5 and recalculate the number of luminaires using the modified *UF*(F) obtained.
7.   Check that the proposed layout does not exceed the maximum *SHR*.
8.   Finally, calculate the illuminance that will be achieved by the final layout.

The values of illuminance required should be above the minimum values for offshore locations shown in Table 10.2.

Table 10.2 Minimum illuminance values offshore

Illuminance (lux)	*Location*
50	External areas (floodlit)
75	External walkways
100	Normally unmanned modules
100	Escape routes (normal lighting)
0.2 (minimum)	Escape routes (escape lighting only)
10–30	Illumination of safety and fire fighting controls (escape lighting only)
100	Unmanned switchrooms
100	Manned areas and accommodation, except:
500	Galleys
300	Laundries and offices
50	Cabins
150	Bedheads

10.2.2 Illuminance at a point

The majority of situations offshore demand other methods of illuminance calculation, since congested or odd shaped areas and specific lighting of structures will render average illuminance calculations inadequate or meaningless.

In such circumstances it is necessary to calculate the illuminance at particular points using one of the following:

Basic photometric data   Basic photometric data can be used in conjunction with Equations (10.1) to (10.7) to calculate the illuminance at particular points by hand. If the number of point calculations are few, this method will usually provide sufficient information for the selection and positioning of several luminaires. It will become laborious and time consuming if a significant number of points and/or luminaires are involved.

Precalculated manufacturer's designs aids, such as isolux diagrams   Isolux diagrams, which show contours of equal illuminance on a specified plane, offer a faster method of performing these calculations. A typical isolux diagram is shown in Figure 10.5. The calculation process can be accelerated using a spreadsheet computer program for the repetitive calculations.

Specific computer programs   Although some limited design programs do exist, most programs simulate the illuminance pattern produced by a chosen layout of luminaires. Point-by-point calculations are required particularly for floodlighting schemes in order to examine the uniformity of illuminance. A program is available for such calculations from Andrew Chalmers and Mitchell Ltd, called Chalmit Litescheme.

10.3   Floodlighting

General information on floodlighting can be obtained from the CIBSE Technical Report TR13 *Industrial Area Floodlighting*.

In floodlighting, the effective source intensity is the photometric value multiplied by the maintenance factor and by the atmospheric light loss factor. The most useful form of luminaire intensity data uses the vertical/horizontal (V/H) coordinate system, such as the isocandela contours shown in Figure 10.6. This accompanies a zonal flux diagram deliberately, for ease of point-by-point calculation after a lumen calculation.

Equation (10.1) can be rewritten as

$$E_H = \frac{I\cos^3\theta}{H^2}$$

where E_H is the horizontal illuminance due to a source mounted H metres above the horizontal.

When an external area such as a drillpipe laydown area is to be lit, the contribution of each floodlight installed can be summed at the points of minimum intensity (usually the corners) and the minimum value of illuminance found. The change in illuminance over the working area to be lit is a measure of uniformity, and is quoted in CIBSE TM5 in terms of the following:

TEST 1

AREA 1

Date: 28 January 1992

Scheme: G MPD0998

Ref: 09980006/ 1

AIMING DATA

AIM No	TOWER X Pos	TOWER Y Pos	TOWER Z Pos	XAIM	YAIM	ELEV deg	HORZ deg	FLUX klm	No Flds	CAT No
1	15.0	25.0	8.0	15.0	10.0	61.9	90.0	28.0	1	OT 250

Illuminance: On Horizontal Plane

Grid: Horizontal at Z = 0

30.00	0	0	0	0	0	0	0	0	0	0	0	0
27.50	0	0	0	1	1	2	3	2	1	1	0	0
25.00	0	1	4	10	23	43	54	43	23	10	4	1
22.50	1	3	7	14	30	50	59	50	30	14	7	3
20.00	2	4	8	16	28	39	44	39	28	16	8	4
17.50	3	5	11	21	35	48	54	48	35	21	11	5
15.00	3	8	16	29	44	59	66	59	44	29	16	8
12.50	5	10	19	31	43	54	58	54	43	31	19	10
10.00	6	11	19	26	34	40	43	40	34	26	19	11
7.50	6	10	14	18	22	25	26	25	22	18	14	10
5.00	5	8	10	12	14	16	16	16	14	12	10	8
2.50	4	6	7	8	9	10	10	10	9	8	7	6
.00	3	4	5	5	6	6	6	6	6	5	5	4

.0   2.5   5.0   7.5 10.0 12.5 15.0 17.5 20.0 22.5 25.0 27.5

x direction ------>   Dimensions in m

Minimum illuminance:	0 lux	Utilisation factor:	0.48
Mean illuminance:	16 lux	Uniformity (Min/Max):	0.00
Maximum illuminance:	66 lux	Uniformity (Min/Avg):	0.00

Note:- Aiming direction and location of floodlight shown by ▽

Figure 10.5 Illuminance chart produced by computer for the floodlight in Figure 10.6. (Courtesy of Thorn Lighting Ltd., Borehamwood)

Criterion	Value
Maximum to minimum	20:1
Average to minimum	10:1
Minimum distance over which a 20% change in illuminance occurs	2 m

10.4   Accommodation lighting

10.4.1 Normal lighting

It is not normally possible to provide lighting of the quality and flexibility of a good hotel, but it is important that the normal lighting is restful and

SONPAK

TLL000502

Description: General area floodlight

Version	Catalogue Number	Lamp
1	OT 250	250W SON-T

Beam Data

Peak intensity(X) cd/klm	: 940	
Beam factor to 10% peak	: 0.68	
Beam Angle to 10% peak	Horizontal: 2x51°	
	Vertical : 43°/66°	
Beam Angle to 50% peak	Horizontal: 2x36°	
	Vertical : 11°/14°	
Beam Angle to 1% peak	Horizontal: 2x67°	
	Vertical : 50°/79°	

(θ) = 22°

Isocandela and Zonal Flux Diagram

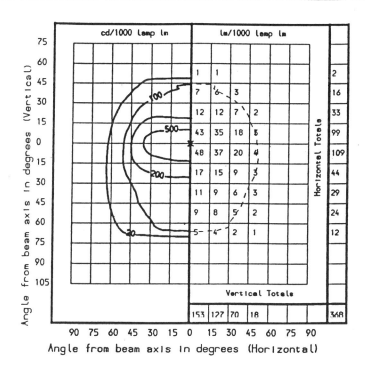

X-indicates the position of peak intensity. The pecked line shows 10% peak contour.
The direction of the peak intensity in the vertical plane is 22 degrees above the normal to the front face of the floodlight.

Figure 10.6 Isocandela and zonal flux diagram for hazardous area floodlight. (Courtesy of Thorn Lighting Ltd., Borehamwood)

blends with a hopefully pleasant decor. Supplementary lighting will be required for reading and writing areas. Where televisions are viewed, care must be taken to prevent distracting reflections appearing on the screen.

Galleys require a much higher level of lighting with good colour rendering for safety and hygiene.

It is usual in the sleeping cabins for each bed to be fitted with a combined lamp and entertainment bedhead unit located conveniently for operation by the bed occupant, such that he or she may read without shadow and without head injury when getting up.

10.4.2 Accommodation emergency lighting

Suitable hazardous area certified emergency lighting must be provided for the adequate lighting of all rooms and escape routes. Each luminaire should be complete with a battery, charger and inverter. Schemes where all the emergency lighting is operated from a single battery are not recommended for reliability reasons.

Each sleeping cabin must be provided with a hazardous area certified emergency luminaire located in such a position as to illuminate the exit route sufficiently to overcome sleep disorientation. Suitable non-industrial designs of certified luminaire with an acceptable appearance for use in accommodation areas are now available.

With the exception of the sleeping cabin luminaires, emergency lighting is left operating continuously. The sleeping cabin units are provided with a mains failure sensing device and are switched on automatically on mains failure.

10.5   Process area lighting

To operate satisfactorily and safely in hazardous process areas, luminaires are required to be mechanically strong in order to resist accidental blows from passing equipment, tools or scaffolding poles. Plastic cases or lenses must not crack due to vibration fatigue, and the casing should be as free from crevices as possible to prevent the trapping of dirt, oil etc. and for ease of cleaning.

The luminaire must be certified for use in hazardous areas, and is usually of the Ex'e' increased safety type. If luminaires are not certified by the British Approvals Service for Electrical Equipment in Flammable Atmospheres (BASEEFA) then it should be checked that the Ex'e' certification is to European Norms, denoted by the certification code EEx'e'. Some Ex'e' certification does not meet the relevant CENELEC/Euronorm standards and may be rejected by the operating oil company. Preferably the luminaire should incorporate a terminal box such that the luminaire enclosure remains sealed when terminals are accessed; otherwise a separate looping-in terminal box will be required adjacent to each luminaire.

Problems have been experienced with Ex'e' luminaires owing to water ingress, possibly leading to arcing. One of the possible scenarios for this is as follows:

1. Gas is detected in a hazardous module.
2. The ESD system initiates a shutdown of the affected plant.
3. As a consequence of this shutdown, either directly owing to the shutdown logic or indirectly owing to fuel gas starvation, for example, the main generation shuts down.
4. With no power available, the air compressors stop, leaving air operated devices dependent on their reservoirs.
5. One by one, as each air reservoir is depleted, the deluge sets release sea water in the process areas they cover.
6. Water enters the luminaires in the module where gas is leaking. Luminaires with integral batteries allocated to escape lighting will still be live. Essential lighting may also be live in this situation, if the emergency generator is too small to run an air compressor.
7. Arcing may now take place within the affected luminaires providing an ignition source for the released gas.

Similar problems have been experienced with other Ex'e' equipment such as junction boxes, but with these the problem can largely be overcome using a layer of Denso tape over joints. Newer designs of luminaire with better water ingress protection are now available, and these combined with good routine maintenance should reduce the ignition risk. Another possible improvement is to provide fire and gas system logic which shuts down all non-essential electrical equipment within the affected module should high-level gas be detected *and* deluge release occur. Escape lighting would still need to remain energized, however.

In congested areas, floodlighting may be required in order to provide effective lighting, to avoid contamination from chemicals and oil and consequently the need for constant cleaning, and also to avoid obstructing the operator's line of sight with the luminaire itself.

10.6 Drilling areas

It is likely that the most difficult lighting conditions on the offshore installation occur in the vicinity of the drilling derrick. The actual process of drilling involves large quantities of water and drilling mud, which may be oil based and difficult to clean from the luminaires. Mud returning from the well may be very hot which, apart from reducing visibility by producing a mist, may cause deformity in plastic luminaire mouldings. The drill floor and the monkey board need to be well lit, but owing to the movement of the travelling block, swivel, kelly etc. and the drillpipe itself, it is necessary to carefully select suitable locations for luminaires on the derrick so as to minimize the risk of mechanical damage. Apart from fluorescent luminaires for illuminating access ladders, landings and equipment on the derrick itself, the derrick may be used for positioning drill floor and monkey board floodlights. Designers should remember the discomfort and danger involved in maintaining derrick lighting when selecting the type and number required, as, once installed, the luminaires must be regularly cleaned and relamped. Permanent safety hooped access ladders with caged landings at appropriate intervals must be provided.

If there is a suitable high structure nearby which overlooks the derrick, on which floodlights may be mounted, the above problem may be alleviated to some extent.

Drilling and wellhead areas present a flammable gas hazard, and luminaires within this area must be certified for use in the hazardous zone in which they are located.

10.7 Laydown areas

Because of the continual movement of containers and equipment around these areas, and the crane operations involved, laydown areas should be floodlit rather than locally lit. While illuminance calculations are in progress, it is important to study crane operating loci in order to establish suitable locations for floodlights to ensure that, as well as providing good lighting of the area, they are unlikely to come into contact with parts of any cranes or their moving loads.

10.8 Helidecks

The landing areas should be outlined by alternate yellow and blue lights which must be visible in all directions. The lights should not be below deck level or protrude more than 0.25 metres above deck level. These lights should be spaced at 3 metre intervals round the perimeter. The yellow lights should have an intensity of 15 candelas minimum and the blue lights of at least 5 candelas with working filters and shades fitted. Higher intensities will assist in poor daylight visibility, but circuits should incorporate a brilliance control to enable the intensity to be reduced at night.

Dual fittings should be installed where access for lamp replacement will take some time, owing to the need to erect scaffolding, ladders etc. To allow for the landing of helicopters during emergencies, a secure, battery fed power supply will be required. The supply system should be capable of maintaining the lamps at normal brightness for at least one hour after the failure of the normal mains supply.

To assist pilots to see landing obstacles, these should be lit by omnidirectional red lights. These lights should be installed at suitable locations in order to provide the pilot with visual information on the proximity and height of obstacles which exceed the height of and are close to the landing area or close to the boundaries of the 150 degree approach sector.

For obstacles such as crane jibs, where the highest point of the jib boom exceeds the height of the landing area by more than 15 metres, a red omnidirectional light of not less than 25 or more than 200 candelas should be fitted at this highest point. On flare stacks, a lower installation point will need to be selected, as close to the top as possible without exceeding the capabilities of the lamp to withstand the radiant heat of the flare (the flare may not always be lit). The rest of the obstacle structure must be picked out by the installation of similar omnidirectional lamps at 10 metre intervals down to the level of the landing area, unless obscured by other superstructure.

Because of the importance of keeping these lights regularly maintained and cleaned, some convenient means of accessing them should be provided.

As a more accessible alternative to the additional low-output red lights on fixed tower structures, the structure may be floodlit, provided that care is taken in the positioning of floodlights so as to avoid dazzling the pilot. Floodlighting, however, is likely to impose a greater drain on the emergency battery system.

Any ancillary structure within the 150 degree approach sector, which is remote from but within a distance of 1000 metres of the main installation, should be similarly lit if the structure significantly exceeds the height of the landing area.

The requirements for the lighting of offshore helicopter landing areas are detailed in the Civil Aviation Authority publication CAP437 *Offshore Helicopter Landing Areas: Guidance on Standards*.

10.9 Jacket and leg lighting

It is important that the lower parts of oil platforms and the immediate area of sea around them are lit. The lighting is required for the following reasons:

(a) in case of a man overboard;
(b) to aid in the manoeuvring of supply and service vessels close to the platform;
(c) to aid in the detection of leaks or inadvertent spillages into the sea.

This lighting is usually provided by high-pressure sodium floodlights, located on a lower deck, preferably in a non-hazardous area, positioned so that they may be directed at the jacket and adjacent area of sea.

The floodlights should be contained in corrosion resistant metal (stainless steel or gunmetal) or an impact resistant plastic enclosure. If located in a hazardous area, both the luminaire and its igniter/ballast unit will need to be certified for such use.

10.10 Navigational lighting

It is a statutory requirement that every platform is equipped with navigation lights which may operate independently for 96 hours following loss of platform electrical power. The following details are based on the *Standard Marking Schedule for Offshore Installations* issued by the Department of Trade.

The light system normally consists of two main, two secondary and various subsidiary flashing lights, an electronic controller, a battery charger and a battery of required capacity.

The main lights are operated from the normal platform supply. They operate in unison, exhibiting the morse letter U. The composition of this morse character must conform to the following:

1. The duration of each dot must be equivalent to the duration of darkness between the dots and of that between the dot and the dash.
2. The duration of the dash must be three times the duration of one dot.
3. The darkness interval between successive morse characters must not be less than 8 seconds or more than 12 seconds.

The lights must be arranged to give an unobstructed white light normally visible for 15 nautical miles in any direction. The lights must be mounted at a height of normally less than 30 metres (but exceptionally 35 metres) above mean high water spring (MHWS), and the beam axis must be directed at the horizon.

The apparent intensity of each light assembly, after all losses, should not be less than 14 000 candela. The total beam width in the vertical plane should not be less than 2.5 degrees at the point on the curve of intensity distribution where the intensity is 10% of maximum.

The lamp assembly is normally equipped with a two-filament lamp in a rotating lamp changer unit (charged with up to six lamps) which automatically ensures that a functioning filament is in circuit and in the correct focal position. The failure of the light assembly for any reason should cause an alarm to be initiated and the secondary battery operated lights to be switched on.

The secondary and main lamps are normally positioned one above the other, with the secondary positioned below. The secondary lamp assemblies are of similar design to the main lamp assemblies, but the lamps are less powerful, having a range of 10 nautical miles.

Subsidiary red flashing lights are required at the corners not marked by the main and secondary lights. These have the same characteristics as the secondary lights and are also battery backed, but have a nominal range of 3 nautical miles.

10.11   Walkways, catwalks and stairways

Installation personnel need to be able to make their way to most parts of the platform, at any time of the day or night, and in any weather. Therefore good lighting of all accessways and particularly stairways is vital for their safety. Most if not all installation lighting outside the accommodation areas will remain on continuously day and night.

10.12   Emergency access lighting

Throughout the installation at least one in every four luminaires should be of the integral battery emergency type. These luminaires remain lit at all times, and will provide illumination for at least one hour following a power failure. These lamps should be placed at strategic locations on escape routes such as inside process modules at ladders and exit doors to ensure adequate escape lighting is available in emergencies. They should also illuminate the controls to fixed fire fighting equipment. This may require the one-in-four ratio to be increased, and in congested areas all luminaires may require to be of the integral battery type.

Certain requirements are laid down in the Offshore Installations Regulations 1974 and 1978 and are therefore mandatory. Notably, these requirements include the following: the minimum duration of escape lighting after failure of normal power supplies (8 hours); and adequate lighting of fire fighting equipment controls during emergencies. It is normally assumed that the emergency generator will be the source of power for at least part of this time.

10.13   Routine maintenance

The importance of regular maintenance and testing of all offshore lighting equipment cannot be overstressed. The example schedule given as Figure 9.1 in the CIBSE *Application Guide: Lighting in Hostile and Hazardous Environments* may be used in the absence of a schedule prepared specifically for the installation concerned.

BS 5345: Part 1 provides guidance on inspection intervals, and these should be adhered to. Escape lighting should be regularly tested, and intervals between testing reduced if more than one or two luminaires are found to have failed in any one area. Escape lighting will receive particular attention during certification authority and HSE surveys.

Chapter 11

Subsea supplies

To those new to the subject of offshore electrical engineering, using electrical power on equipment which is totally immersed in highly conducting salt water may seem somewhat foolhardy. In actual fact, electrical power has been used in submersible pumps and for underwater arc welding since around the turn of the century. Providing certain precautions are taken, using electricity under the sea can be comparatively safe.

11.1 Subsea power supplies

Permanent arrangements will be required for power to operate valves, switches and sensors on underwater wellhead manifolds and the new generation of subsea production wells which are serviced periodically by special vessels. Where power and control are required at the manifold from a platform some distance away, an unearthed supply will be needed at a suitable voltage to compensate for the impedance of the subsea cable. It is usual, however, for power for operation of valves etc. to be transmitted hydraulically with electrical actuation from a secure (UPS) AC supply on the adjacent platform.

It is recommended that all AC supplies used are three-phase and are isolated via the delta connected, unearthed secondary windings of transformers. This provides the best scope for incorporating protection devices, and single earth faults can be arranged to operate alarms whilst the circuit remains in operation.

11.2 Diver's life support equipment

The main requirement for the North Sea diver is for heat, both for body and for respiratory gas heaters.

The design of respiratory gas heating, if fitted to the diver's suit, must be such that there is no possibility of electrocution through the mouth. If the diver's suit is heated electrically then the elements of heated undergarments should be screened such that there is no possibility of the heating elements touching the diver's skin, or the elements shorting internally and causing localized heating.

Both respiratory gas heaters and electric body heaters should use a low-voltage DC supply no greater than 24 V DC. Heater elements, including supply cables, should be screened, with the screen continuity continuously monitored.

A line insulation monitor (LIM), designed on failsafe principles, should be incorporated into the supply circuit to detect any deterioration in the supply cable to heater element insulation resistance. The sensitivity of the LIM should be such as to ensure a tripping response in a few milliseconds to a significant fall in resistance.

By far the most common form of diver's suit heating in the North Sea, however, is the hot water machine from which heated sea water is pumped from the surface. The heated water is directly in contact with the diver's skin and escapes from wrist and ankle bands in the suit. There is a remote possibility that an earth leakage current could travel down the heated water via the hose and be earthed via the diver. However, if the heater element tank in the hot water machine is adequately earthed, there should be no danger, especially as the incoming cold water also provides a conducting path.

If the diver is to use any hand held electrical equipment, sensitive earth leakage circuit breaker (ELCB) protection will be required. ELCBs with a trip current of 4 mA are recommended (see codes of practice in Bibliography).

11.3 Diving chambers

As the lives of divers may depend on the electrical supply being available at the chamber, particularly if deep diving involving compression chambers is being undertaken, the electrical supply systems of the associated support vessels must be of the highest integrity and reliability.

The equipment used inside diving chambers is, where possible, to be designed using the principles of hazardous area intrinsic safety. This is for the following reasons:

1. It will prevent ignition should flammable well gases be present.
2. It will help to prevent fires in a chamber with an oxygen enriched atmosphere.
3. It will ensure earth fault currents are limited to very low values.

Table 11.1 shows the effects on the human body of various magnitudes of AC current. From this it can be seen that values beyond 18 mA are dangerous. In salt water the resistances tend to be low, and very low voltages are required to achieve this threshold value. All electrical circuits within the chamber will need to be protected against earth faults by using current operated earth leakage circuit breakers set to trip at not more than 15 mA in 30 ms.

The power supply within the chamber should be low voltage and either direct current (e.g. 24 V DC) or possibly at a frequency greater than

Table 11.1

Current at 50 Hz to 60 Hz rms milliamperes	Duration of shock	Physiological effects on humans
0−1	Not critical	Range up to threshold of perception, electric current not felt.
1−18	Not critical	Independent release of hands from objects gripped no longer possible. Possibly powerful and sometimes painful effect on the muscles of fingers and arms.
18−30	Minutes	Cramp-like contraction of arms. Difficulty in breathing. Rise in blood pressure. Limit of tolerability.
30−50	Seconds to minutes	Heart irregularities. Rise in blood pressure. Powerful cramp effect. Unconsciousness. Ventricular fibrillation if long shock at upper limit of range.
	Less than cardiac cycle	No ventricular fibrillation, heavy shock.
50 to a few hundred	Above one cardiac cycle	Ventricular fibrillation. Beginning of electrocution in relation to heart phase not important. (Disturbance of stimulus conducting system?) Unconsciousness. Current marks.
	Less than cardiac cycle	Ventricular fibrillation. Beginning of electrocution in relation to heart phase important. Initiation of fibrillation only in the sensitive phase. (Direct stimulatory effect on heart muscle?) Unconsciousness. Current marks.
Above a few hundred	Over one cardiac cycle	Reversible cardiac arrest. Range of electrical defibrillation. Unconsciousness. Current marks. Burns.

20 kHz, to which, apparently, the human body responds in a similar way to direct current. Any sockets used should be the underwater type, which can be connected while immersed in sea water without hazard. Unused sockets should be provided with insulating caps. If no suitable intrinsically safe equipment can be obtained, equipment incorporating other types of protection may be used. Any electric motors in the chamber should be of the flameproof or increased safety type, and designed to withstand stalling currents for long periods without detriment.

It is more than likely that surface Ex'd' certification will prove invalid for equipment to be used in underwater chambers, because of the higher breathing atmosphere pressures required for long-term compression diving. The equipment enclosure will require pressure testing to 1.5 times the normal peak explosion pressure times the working pressure of the chamber (bar A). Intrinsically safe equipment safety margins are also reduced by increase in pressure, and such systems will also require reassessment when used in this environment. Other types of certified equipment may be affected by the higher humidity or by water ingress because of the higher pressures likely to be experienced by underwater equipment.

11.4   Inductive couplers

If the core laminations of a transformer are cut in half, the primary coil wound on one half and the secondary wound on the other, then when the two halves are mated together a rather inefficient transformer is formed. This principle is used in thousands of inductive couplers throughout the North Sea and elsewhere.

Figure 11.1 shows the mating sides of inductive couplers. The core and windings are encapsulated so that the sea water at North Sea sea bed pressures cannot penetrate into the live parts of the device. Only the surface of the core, which is ground flat, is exposed. Because of the extra losses produced by the position of the windings and the air gap in the core, more heat is generated than in a conventional transformer, and this must be removed by a core heat sink connected to the stainless steel housing.

Figure 11.1 Inductive coupling units. (Courtesy of Ferranti Subsea Systems Ltd.)

11.5   Subsea umbilicals and power cables

An umbilical is a set of hydraulic hoses and signal, control and power cables bundled together, armoured and sheathed to form a single entity and used to provide services from a surface installation or vessel to equipment below the surface, often on the sea bed. Individual cables within the umbilical may also be screened or armoured or both, and the whole umbilical may be strengthened against tension damage during deployment by the provision of an embedded steel tension cable.

A subsea cable is usually of simpler construction, containing the three main power cores operating at usually between 11 and 33 kV, with telecommunications and protection pilot wires often located in the centre of the cross-section. The cable will be armoured and sheathed overall and may also contain a tension wire. Cables are brought on to a platform via a J-tube, which is a length of steel guide pipe fixed at various points to the installation jacket or hull and curling out at the bottom to receive the cable from the sea bed.

Whether a subsea link is required between two or more platform electrical power systems, or between a surface installation and sea bed equipment, the need for both a very high level of manufacturing quality control and strict adherence to the planned delivery dates is vital. Several very expensive mistakes have occurred in recent years, where umbilicals of 10 kilometres or more have had to be scrapped because of the discovery after completion that vital cables within the umbilicals were faulty. Apart from the financial penalties involved, the time delays in reconstructing the umbilicals threatened to delay their laying beyond the North Sea summer weather window.

The thorough quality control procedures adopted during construction should include the following:

1. Discussions should be held with the manufacturers in order to arrive at the optimum cable construction for the particular application. This will include:
    (a) length of umbilical;
    (b) depth of water and hence pressure;
    (c) type of cable laying equipment used, handling conditions and tension likely to be applied;
    (d) electrical parameters required to ensure satisfactory operation of the equipment at either end, which as a minimum should include voltage, current, resistance, reactance, capacitance and attenuation over an agreed frequency range;
    (e) storage conditions.
2. Analysis should be made of samples of all materials used for the minimum performance required. Insulation thicknesses should be more generous than those for cables used on land. If screen drain wires are required, ensure that these do not have a cheesewire effect on insulation when the umbilical is in tension.
3. Close control of conductor and insulation diameter and thickness of each cable should be maintained during manufacture.

4. Samples of each cable should be taken at the start and finish of each stage of the cable manufacturing run for analysis, and the results of analysis should be known before continuing with the next stage.
5. The final testing of each cable should include insulation resistance, high-voltage testing and measurement of total cable resistance, reactance, capacitance and attenuation over a given frequency range. If DC switching or digital signals are being transmitted, low attenuation at megahertz frequencies may be important in order to avoid degradation of the signal over the length of the cable. Such readings may also detect faults, if attenuation readings for known healthy cables are available for comparison.
6. A sample of the completed cable should undergo high-voltage testing in water at the pressure equivalent to at least the maximum depth of water likely in the installed location. The test is carried out using a pressure vessel with a cable gland at each end. The cable sample is passed between the two glands, which are then sealed. The vessel is filled and pressurized and the high voltage can then be applied at the cable ends protruding from the glands.
7. Another high voltage test often conducted on a sample is to submerge the sample in a tray of salt water with one exposed conductor in air at either end. A high voltage is then applied between a submerged conductor (or the tray) and the exposed conductor in air. This and the previous test are carried out to the maximum voltage available or until a voltage breakdown occurs in the sample.
8. Once all the cables have successfully completed their testing, they may be incorporated into the umbilical. However, if there is a delay between cable and umbilical manufacture, or the cable has to be transported to a different site, further insulation testing will be required immediately prior to umbilical manufacture to ensure that transit or coiling damage has not occurred. It should also be remembered that faults due to poor curing of the (plastic) insulation may appear several weeks after manufacture.

11.6 Cathodic protection

Cathodic protection provides an effective method of mitigating corrosion damage to subsea metal structures, whether their surfaces are coated or not.

Impressed current cathodic protection is preferred for long term protection of both platform structures and subsea pipelines whereas galvanic or sacrificial anode systems are normally only used where power supplies are unavailable, or for temporary protection during construction, tow-out and commissioning. A galvanic anode system may also be used to complement an impressed current system in order to achieve rapid polarization and as part of an overall design to protect the structure.

Although there are guides and codes of practice (such as the one listed in the Bibliography) to assist the engineer, the design of a good system depends to a large extent on past experience with similar structures and subsea environments.

11.7   Types of system

11.7.1   *Impressed current systems*

In this type of system, an external source of power, usually a transformer-rectifier unit, provides the driving voltage between the anode and the structure to be protected. The negative terminal of the DC source is electrically connected to the structure and the positive terminal similarly connected to the anode. The output voltage of the DC source is adjustable so that the protection current may be varied either by a trimming potentiometer or automatically through a control signal loop to provide the required voltage at the reference half-cell. Transformer-rectifiers for this use are usually rated in the range of 20–500 A and from 30 to 120 V.

Impressed current anodes, unlike galvanic ones, must be well insulated from the protected structure if they are to be mounted on it. Where it is found necessary to place impressed current anodes very close to the protected structure, dielectric shielding between anode and structure must be provided.

As it has a uniformly low electrical resistivity, sea water is an excellent medium for the application of cathodic protection and facilitates an even current distribution over the surface to be protected. Bare steel submerged in sea water is easily polarized if an adequate current density is maintained.

Impressed current anodes may be of silicon cast iron, lead alloy, or graphite. Lead silver alloy or platinized anodes, although more expensive at the installation stage, may be used to provide a longer life and hence maintenance cost savings. Impressed current anode systems should be designed for a life of at least 10 years. High silicon cast iron corrodes relatively slowly and in sea water, where chlorine is produced at the anode surface, chromium may be added to further improve longevity.

The resistance of a single anode in free flowing sea water is less than that of the same anode in mud or silt, and it may be necessary to install groups of anodes, sometimes mounted on a wood or concrete framework, in order to reduce silting up and maintain effective contact with the sea water. Care must be taken to avoid incidental rubbing or scraping of the active anode surface by suspension or stabilizing rods or cables, since the effect of this is to accelerate dissolution at that point. For obvious reasons, suspending the anode by the conducting cable is not advised. The number of anodes is determined by the required anode life, the allowable current density, and the circuit resistance of the system. Anode life will be affected by the operating current density and the total current magnitude. In free flowing sea water, a maximum output of about 10 A m^{-2} is normal practice, giving an anode consumption rate in the region of 0.4 kg per amp per year.

Platinized anodes are generally constructed from a solid or copper-cored lead alloy rod about 12 mm in diameter and 1500 mm long. A layer of platinum approximately 0.005 mm thick is deposited on this. A similar thickness of platinum may be used on anodes of other shapes and other materials, such as platinized titanium, platinized niobium or lead silver alloy. Graphite is not generally suitable for use offshore.

For reasons of economy, platinized titanium anodes should not be used where silt or mud may accumulate, above the lowest tide level, or where fluctuating voltages, single (i.e. low frequency) AC ripple or anode voltages higher than 7 volts are present, since the expected long life will not materialize in these conditions. The system design should be such that these anodes are not expected to carry current densities of more than 700 A m^{-2}. The 0.005 mm thick platinizing dissolves at the rate of about 10 milligrams per ampere year, and although the abrasive action of water-bourne sand may reduce this to some degree, a life expectancy of 15 years may be realistically hoped for.

The anodes are usually mounted on supports cantilevered out, but electrically insulated from the protected structure. In shallow waters, seabed non-metallic frames may be used to support the anodes above the mud or silt level. To protect any offshore platform, distributed multiple anode arrangements will be required to be mounted over the members of the complex subsea structure to provide a uniform current density. The size and design of the cathodic protection scheme will depend on the current density required to bring the structure up to the level of protection required. Sea water velocity and oxygen content will affect chemical activity and hence the current density. Statistical data will be required from the platform site, if an accurate figure for current density is to be determined. However, data may be available for similar environments as a guide. Some figures are given below as a indication of the magnitude required:

Location	Environment	Required current (mA m^{-2})
Persian Gulf	Sea water	20
	Below mud level	50
North Sea	Sea water	50
	Below mud level	90

It should be remembered that there will be little or no protection afforded by these systems above the low tide level and additional forms of protection will be required above this level.

11.7.2 Impressed current systems on submerged pipelines

Offshore submerged pipelines generally require a minimum pipe-to-water potential of -0.9 (ON) V with reference to a silver/silver chloride half cell. This corresponds to a current density of 2.5 mA m^{-2}.

11.7.3 Galvanic anode systems

In the galvanic or sacrificial anode system, the driving voltage between the structure to be protected and the anodes is developed directly by the electrolytic potential between the two (different) metals involved. The anodes are usually of aluminium alloy or magnesium, types which are less affected by insulation from oil wetting such as 'Galvalum III' being often preferred. Galvanic anodes may be placed very close to, or in contact with, the structure to be protected. Galvanic anode systems should be designed for a life of 25 years.

'Galvanum III' or similar anodes are suitable for use on both subsea pipelines and offshore structures and are particularly suited for use in saline muds or silted-over conditions. On subsea pipelines it is used in the form of 'bracelets'. On subsea structures, bar type anodes are mounted at suitable intervals.

Although used extensively on ships, zinc and magnesium anodes are not suitable for offshore use mainly due to their faster rate of consumption and hence higher costs involved in replacement. Supplementary magnesium anodes may be used in order to provide a temporary boost in polarizing current, however. As cathodic protection starts to operate, a layer of alkaline material is formed on the protected structure by cathodic electrode reaction. Provided this material is not dislodged, its presence reduces the current density required to maintain protection.

On offshore structures the polarization-boost anode may take the form of a 10 × 20 mm ribbon of magnesium attached to the structure and supplementing the permanent system.

11.8   Cathodic protection calculations

This section is for guidance only and results should be heavily weighted by previous experience and measured data from previous structures and tests carried out for the new structure to be protected.

Impressed current circuits basically consist of a DC power source driving a current through the anode, the electrolyte (i.e. sea water), and back through the structure. The empirical rule used for calculating the current is:

$$\text{Total current (A)} = \frac{\text{Surface area (m}^2) \times \text{mA m}^{-2}}{1000}$$

The surface area used is that of all metal surfaces submerged in the electrolyte at mean high water.

Having determined the total current, the component resistances.

Cable resistance (both poles)	R_C
Anode to electrolyte resistance	R_A
Cathode (structure) to electrolyte resistance	R_E
Linear resistance of the structure	R_S

are adjusted in order to match the total DC output of the available power source.

The linear resistance of an offshore platform may be neglected in most cases, but where length is extensive in comparison with cross-section as in subsea pipelines, R_S should be evaluated.

A similar calculation is required for galvanic anodes as follows:

$$I_a = \frac{E_d}{R_a} = \frac{E_a - E_{cu}}{R_a} \text{ A}$$

where: I_a = Anode current output (A)
R_a = Anode to electrolyte resistance (Ω)
F_a = Open circuit potential between anode and electrolyte, measured by reference electrode (V)
E_{cu} = Structure to electrolyte potential when fully protected (V)
E_d = Driving voltage between anode and protected structure (V) = $E_a - E_{cu} = V_s + V_a$

Where V_s = structure to electrolyte potential and
V_a = anode to electrolyte potential

Typical values of E_a are:

Galvomag	1.7 V
Magnesium	1.5 V
Zinc	1.05 V
'Galvanum III'	1.05 V

Calculation of anode consumption

The expected life of an anode in years is given by:

$$Y = \frac{CR \times W}{8760 \times A}$$

where: Y = Life (years)
CR = Consumption rate (Ah kg^{-1})
W = Anode weight (kg)
A = Anode output (A)

A typical value of CR for magnesium is 1230 Ah kg^{-1}.

Chapter 12

Reliability

This chapter has been included as an introduction to the use of reliability analysis. This aids the offshore electrical system designer in his quest to produce a system that is not only fit for its purpose when all its component parts are working, but will continue to function in some fashion, dependent on the severity of the problem, when certain of its components fail or incidents occur on the installation which affect the system's integrity. Reliability analysis encompasses a number of graphical, mathematical and textual operations which present the known facts, statistical data and/or experience about the proposed system or similar systems in a way which highlights its weaknesses or ranks the effectiveness of the options available to the designer.

It is very important when carrying out any form of reliability analysis that specific goals are selected. The availability of an electrical supply at a particular point in a system will be very much dependent on the configuration of the system and where that point is located in it. Therefore a vague instruction such as 'quantify the reliability of the supply system' is very unhelpful, and it would be necessary to discuss with the client concerned what particular supplies were of particular concern and then to consider each in turn. For example, if the object was to maintain full production then only the equipment required to maintain supplies to the production switchboards would need to be considered. However, if supplies to essential equipment only were being analysed, the availability of all generation and all the routes which could be used to connect it to the essential equipment would need to be taken into account and, hopefully, in comparison a much higher figure for the availability of supply would be obtained.

12.1 Duplication and redundancy

Provided that good quality assurance procedures are adhered to throughout a design and construction project, to the extent that all equipment purchased for installation on the platform is fit for its purpose, then theoretically there should be no benefit to be gained by the replacement of one item of equipment by one of similar function from another manufacturer. In practice this is not the case, as we know by experience that certain devices are better obtained from a particular company. For obvious reasons, information of this kind in the form of usable data is difficult to obtain, and therefore for the purposes of this chapter it will be assumed that every component with the same function has the same reliability properties, i.e. its failure rate and its mean time to repair (MTTR) are the same as any other with that particular function. Nevertheless, it is advisable to keep records of failure and times to repair of equipment installed, as analysis of this data in itself may show up some problem when the records are compared with available generic data.

Very significant reliability improvements can be made in systems, usually much more significant than from simply replacing components, by reconfiguring them in some way. Some examples of this are given in Chapter 2. The basic principle for any reconfiguration is to provide redundancy in such a way that, during a component's outage, its function will be maintained by one or more other components of similar function, whether this outage is planned or unplanned. In most cases the component is a system in its own right, such as a generator package with its associated auxiliaries or a subsea cable supply from another platform.

12.2 Failure mode, effects and criticality analysis (FMECA)

This procedure, like hazard and operability analysis (hazop) which is beyond the scope of this book, is a method of identifying all modes of failure, analysing their effects and, where possible, evaluating the frequency of their occurrence. The basic FMECA process is the completion of a table with column headings similar to Figure 12.1.

However, before the columns can be completed, it is necessary to draw a block schematic diagram of the system to be analysed at the level of detail required. This level should be low at the initial stage; as it becomes more obvious which components or subsystems are critical, particular areas can be broken down into more detail. At each level a more detailed block diagram should be drawn.

If we take a circuit breaker as an example item on an FMECA, this would appear as a block in the diagram as shown in Figure 12.2. A circuit breaker is a particularly difficult component to describe in an FMECA, as it has several functions and several failure modes.

The circuit breaker functions are as follows:

1. It connects and isolates an electrical circuit as required.
2. In the event of a fault, it interrupts the flow of fault current automatically in conjunction with a sensing device.

Note that the adequacy of load and fault rating is assumed, as the purpose of this analysis is not to question the system design calculations but to consider the effects on the system of a particular mode of failure.

The circuit breaker has two modes of failure associated with the first function, namely that it:

System: Generation	Failure mode and effect analysis worksheet	Page 1 of 8
Subsystem: Main HV generation	Job no:	Issue A
Drawing no:	Job title: Reliability Study	Prepared by: ___ Date: ___
	Client	Checked by: ___ Date: ___

| (1) Item code | (2) Component | (3) Function | (4) Failure mode | Failure effect | | (7) Means of detection | (8) Severity of effect | (9) Compensating provisions or measures in operation | (10) Remedial measures in design |
				(5) Local level	(6) Plant level				
1	Generator UCP	a) Trip/ Inhibit Start under fault or pre-set conditions	Unit starts under fault conditions	Damage to unit	Down time due to original fault extended. Reduced production	Operator	II	ST/BY unit	Instrumentation and controls, plus maintenance procedures and periods seem adequate from information available
			Unit fails to trip under fault conditions	Damage to unit	Initially no effect. Severe unit damage results in loss of electrical output to distribution reduced production	Operator	II	ST/BY GT Unit if available Load shed	" "
			Unit trips under healthy conditions	Loss of output	Loss of output to distribution load shed. Initiate	Operator	I	" "	" "
		b) Auto sequence control brings set on load through pre-set conditions	Wrong sequence	Damage to unit	" "	Operator	II	Second GT unit	" "
				Fire possible	Fire hazard	Operator/ Control system	III	" "	" "

Figure 12.1 Sheet for use with failure mode, effects and criticality analysis

213

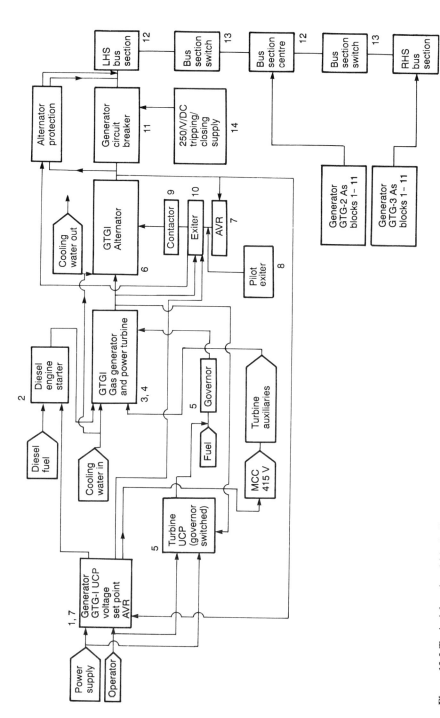

Figure 12.2 Typical functional block diagram for use with FMECAs

(a)   fails to open when isolation/disconnection is required;
(b)   fails to close when power supply is required.

For the fault clearing function there is one failure mode:

(c)   fails to interrupt fault current.

If required this could be split into:

(c)   (i)   fails to interrupt make fault;
(c)   (ii) fails to interrupt break fault.

Lastly, there are faults associated with the breakdown of insulation within the circuit breaker cubicle:

(d)   interphase fault in circuit breaker cubicle;
(e)   earth fault in circuit breaker cubicle.

Some of the faults listed may be due to the failure of the trip/close battery supply or a protection relay. Depending on the level of detail required, however, these components can be included as part of the circuit breaker block, or if necessary a new block diagram may be drawn showing individual blocks for each component in the circuit breaking system.

The failure modes having been identified, the next stage is to describe

(a)   the effect on the system;
(b)   the effect on adjacent equipment;
(c)   the effect on the offshore installation.

If the system is well designed, there should be no effects on the installation and only one or two effects on adjacent equipment, usually associated with fire. All effects listed should have some compensating provision or remedial measure available to compensate, and this should be listed as shown in Figure 12.1 in column 9. Where no such provision or remedy exists, column 10 should be used to propose a solution which prevents such an effect or alleviates the problem. At the end of the study all such proposals should be listed.

If a significant number of proposals appear in column 10, it will probably be necessary to rank the associated effects in order of criticality. This is usually done by drawing a *criticality matrix* as shown in Figure 12.3 (extracted from MIL-STD-1629A). This is a graph with axes of severity of effect against the probability of its occurrence. The severity is given four arbitrary categories as follows:

Category I: catastrophic   A failure which may endanger the whole installation and involve loss of lives.
Category II: critical   A failure which leads to prolonged loss of production and may cause severe injury and major equipment and/or system damage.
Category III: marginal   A failure which may cause minor injury or minor system/equipment damage and will result in delays to or partial loss of production.
Category IV: minor   A failure which is not serious enough to cause injury or any equipment/system damage but which will result in an unscheduled outage for maintenance or repair.

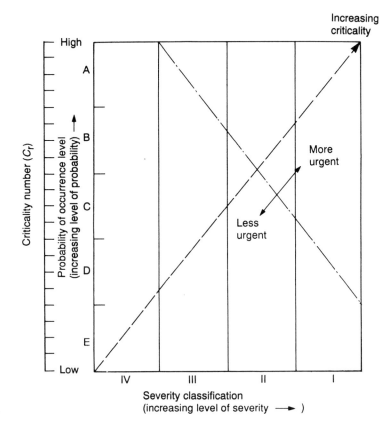

Figure 12.3 Typical criticality matrix

The frequency of occurrence may either be a numerical value obtained from a reliability databank, or a qualitative level assigned as follows:

Level A: frequent   A high probability of occurrence during the time the system is running.
Level B: reasonably probable   A moderate probability of occurrence during the time the system is running.
Level C: occasional   An occasional probability of occurrence during the time the system is running.
Level D: remote   An unlikely probability of occurrence during the time the system is running.
Level E: extremely unlikely   An essentially zero probability of occurrence during the time the system is running.

A diagonal line is drawn on the matrix, and each failure mode effect which has no existing compensating provision should be represented by a cross drawn on the matrix. Failure effects appearing in the upper right hand area of the graph have the greatest criticality and hence the most urgent need for corrective action.

12.3   Fault trees

A fault tree is a graphical method of describing how faults in system components relate to overall system failures. The rules for constructing a fault tree will, if rigidly adhered to, lead to the creation of a graphical model of the system which will greatly assist the system designer in ensuring that as many failure modes as possible are remedied at the design stage.

A typical fault tree model is shown in Figure 12.4. The diamond or circle shapes represent *basic events*, which are component failure modes. The rectangles represent *combination events*, which are the logical result of a combination of basic events. The logic of the tree is shown by using mainly AND and OR gates. The AND gate represents a situation where a combination event can only exist if the basic events connected to it by the AND gate all exist simultaneously. The OR gate represents the situation where a combination event can exist if one or more of the basic events connected to it by the OR gate exist. Other logic devices can be used to represent standby equipment, voting systems etc., and in most cases normal Boolean algebra rules apply.

12.3.1   Fault tree construction

As with all methods of reliability analysis, the system boundaries must be carefully defined. A block diagram of the system should be drawn which shows all external inputs, as these should be identified in the fault tree construction. This process assists the analyst to understand the function of the system. If necessary an FMECA should be undertaken at this stage if the failure modes of the various components are not clear. If complex logic is involved, truth tables should also be constructed.

The top event of the tree must be carefully defined, as an ambiguous or vague title will confuse and lead to much wasted effort. When the top event has been stated, each combination event is defined by asking the following questions:

1.   What events taken singly will directly result in the event under consideration? These events are connected to the event under consideration by an OR gate.
2.   What combination of events will directly result in the event under consideration? These events are connected to the event under consideration by an AND gate.

By definition, all combination events will be the consequence of events drawn below them and connected in some way via logic gates. The fault tree is complete when all combination events are shown to be the cause of two or more basic events.

When the fault tree has been drawn, various methods can be applied to it to identify problems in the system that it represents as follows.

12.3.2 Common mode failure identification

If the same event appears in several places on the fault tree, common mode failures may be identified. Some examples of common mode failures are as follows:

(a) fire in a switchroom, leading to the loss of two or more distribution routes;
(b) fuel gas system fault, leading to all main generators failing;
(c) loss of hazardous area ventilation, leading to shutdown of all main oil line pumps and/or all gas compressors.

12.3.3 Qualitative analysis

This can be carried out by inspection in a simple fault tree by identifying any *minimal cut sets*. These sets consist of a group of basic events which will cause the top event to occur if and only if they exist simultaneously.

Boolean reduction may also be used provided there are no exotic logic gates such as those with a sequential or timed inhibit function.

12.3.4 Quantitative analysis

The two main methods of analysing a fault tree quantitatively are as follows.

12.3.4.1 Evaluation using event probabilities

This method is used when the event failure data is available in terms of probabilities.

Where two events are output through an AND gate (see Figure 12.5), the failure probabilities may be multiplied as follows:

$$P(a.b) = P(a) P(b) \quad \text{where } (a.b) \text{ represents } a \text{ AND } b$$

Where two events are output through an OR gate, the resultant probability of failure is as follows:

$$P(a+b) = P(a) + P(b) - P(a) P(b) \quad \text{where } (a + b) \text{ represents } a \text{ OR } b$$

If the probabilities $P(a)$ and $P(b)$ are each 0.1 or less, then the product $P(a) P(b)$ is very small compared with the sum $P(a) + P(b)$, and a good approximation may be obtained from

$$P(a+b) = P(a) + P(b)$$

This may be proved for n events as follows (Green and Bourne 1972):

$$P(a.b. \ ... \ .n) = P(a) P(b) \ ...P(n) \qquad \text{for AND gates}$$
$$P(a+b+ \ ... \ +n) = P(a) + P(b) + \ ...+ P(n) \qquad \text{for OR gates}$$

Note that the above are only true for *independent events*.

218

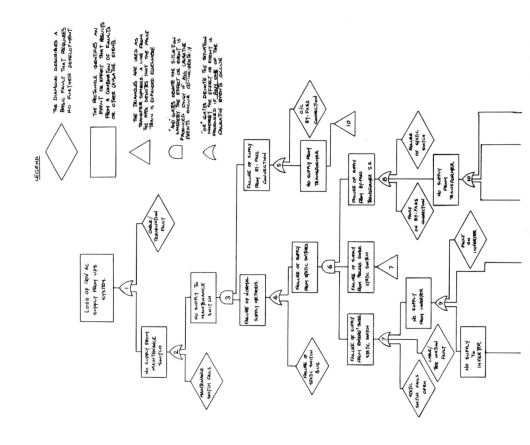

LEGEND

THE DIAMOND DESCRIBES A BASIC FAULT THAT REQUIRES NO FURTHER DEVELOPMENT

THE RECTANGLE IDENTIFIES AN EVENT OR EFFECT THAT RESULTS FROM A COMBINATION OF FAULTS OR OTHER CAUSATIVE EVENTS.

THE TRIANGLES ARE USED AS TRANSFER SYMBOLS. A LINE FROM THE APEX DENOTES THAT THE FAULT TRAIN IS EXPANDED ELSEWHERE

"AND" GATES DENOTE THE SITUATION WHEREBY THE EFFECT OR EVENT IS PRODUCED ONLY IF ALL CAUSATIVE EVENTS OCCUR SIMULTANEOUSLY

"OR" GATES DENOTE THE SITUATION WHEREBY THE EFFECT OR EVENT IS PRODUCED IF ANY ONE OF THE CAUSATIVE EVENTS OCCUR

LOSS OF 110V AC SUPPLY FROM UPS SYSTEM.

CABLE/ TERMINATION FAULT

NO SUPPLY FROM MAINTENANCE SWITCH — 2

MAINTENANCE SWITCH FAILS

NO SUPPLY TO MAINTENANCE SWITCH — 3

FAILURE OF NORMAL SUPPLY METHODS. — 4

FAILURE OF SUPPLY FROM BY-PASS CONNECTION — 5

FAILURE OF STATIC SWITCH BUS

FAILURE OF SUPPLY FROM STATIC BY-PASS — 6

O/C BY-PASS CONNECTION

NO SUPPLY FROM TRANSFORMER — 10

FAILURE OF SUPPLY FROM BY-PASS TRANSFORMER S.S.

FAULT ON BY-PASS CONNECTION

FAILURE OF SUPPLY STATIC SWITCH

FAILURE OF SUPPLY FROM PROCESS SWITCH STATIC SWITCH — 7

FAILURE OF SUPPLY FROM ENERGY SWITCH STATIC SWITCH — 7

STATIC SWITCH FAILS OPEN

CABLE TERMINATION FAULT

NO SUPPLY TO INVERTER

NO SUPPLY FROM INVERTER — 9

FAILURE OF STATIC SWITCH

NO SUPPLY FROM TRANSFORMER — 10

FAULT ON INVERTER

219

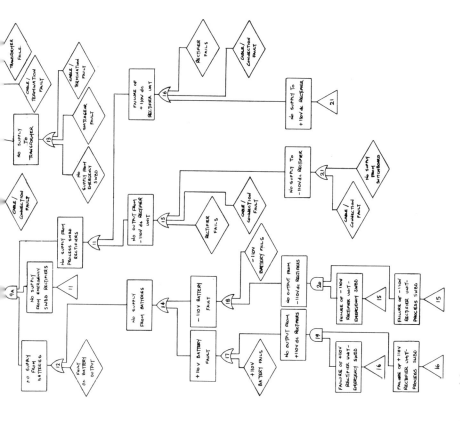

Figure 12.4 Typical fault tree model

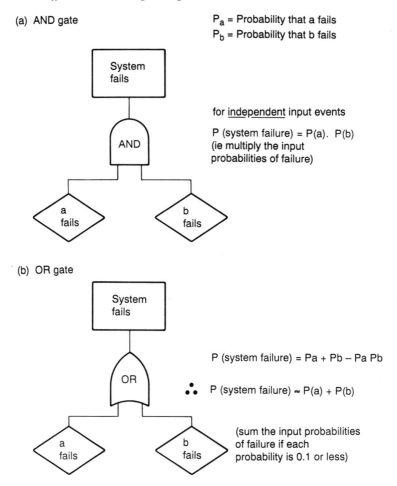

(a) AND gate

P_a = Probability that a fails
P_b = Probability that b fails

System
fails

AND

for <u>independent</u> input events

P (system failure) = P(a). P(b)
(ie multiply the input
probabilities of failure)

a
fails

b
fails

(b) OR gate

System
fails

OR

P (system failure) = Pa + Pb – Pa Pb

∴ P (system failure) ~ P(a) + P(b)

a
fails

b
fails

(sum the input probabilities
of failure if each
probability is 0.1 or less)

Figure 12.5 AND logic for fault tree evaluation

12.3.4.2 Evaluation using event failure rates

For any number of events with constant failure rates input to an OR gate, it can be proved (Green and Bourne 1972) that the output has a constant failure rate which is the sum of the failure rates of the inputs. For any number of events with constant failure rates input to an AND gate, it can be proved that the output failure rate after a given time t will be a function of t. If each of the events is identical, as would be the case with the failure rates for a number of generators in a system where each is capable of maintaining the full system load, then without maintenance the output failure rate would tend to approach the single-unit failure rate after a certain number of hours (see Figure 12.6).

In a real situation, where regular maintenance is carried out, as a good approximation it is acceptable to take the output of an AND gate as the product of the input event failure probability, provided mean times to

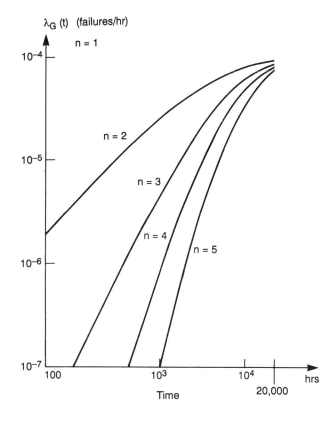

$\lambda_G(t)$ (failures/hr)

Figure 12.6 Graph of system failure rate against unit numbers, without maintenance

repair (MTTRs) are very much shorter than mean times between failures (MTBFs).

12.4   Reliability block diagrams

If a system is broken down into key components or elements, the failure of which would have some effect on the system's availability, a block diagram may be drawn such that, for the continuing operation of the system, a continuous string of elements must exist from one side of the diagram to the other. An example of such a block diagram is shown in Figure 12.7. A reliability block diagram, produced in this way, not only provides a clear indication of any critical system components but can often be numerically evaluated.

In order to carry out this numerical evaluation, it is necessary to assign a value for failure rate and mean time to repair for each system element. An availability A may then be calculated for each element as follows:

$$A = \frac{\mu}{\lambda + \mu}$$

222

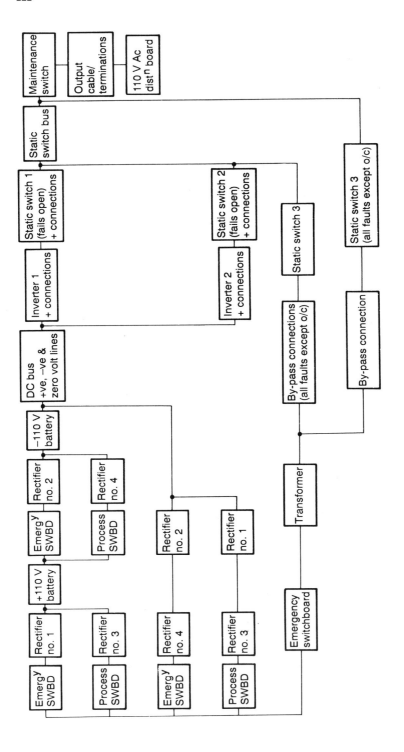

Figure 12.7 Typical reliability block diagram

where μ is the reciprocal of the mean time to repair (hours), and λ is the failure rate (failures per 10^6 hours). The unavailability is given by

$$\overline{A} = 1 - A$$

and the unavailability of a system of n components in series is then

$$\overline{A}_s = \overline{A}_1 + \overline{A}_2 + \dots + \overline{A}_n$$

If the unavailabilities are calculated for each component, they may be added together where the configuration consists of series components and where the component MTBFs are very much larger than their MTTRs. The MTBF can be obtained by inverting the failure rate and multiplying by 10^6 as follows:

$$1/(\lambda \text{ failures}/10^6 \text{ hours}) \equiv 10^6/\lambda \text{ hours/failure}$$

The unavailability of a system of n identical components in parallel, assuming only one component is required to maintain the system, is

$$\overline{A}_s = \overline{A}_1 \overline{A}_2 \dots \overline{A}_n$$

Where more than one parallel component is required to maintain the system, the following can be shown to be the combined availability:

$$A_s = \sum_{j=m}^{n} \binom{n}{j} A^j (1-A)^{n-j}$$

where n is the total number of components, and m is the number of components required to maintain the system.

With parallel components, calculation becomes more complicated. In a system X where two parallel components are redundant, i.e. both are designed to be functioning but only one is required for system operation (Figure 12.10), the unavailability of the system is

$$\overline{A}_X = \overline{A}_A.\overline{A}_B$$

If there are three components in parallel and two out of the three are required for successful system operation, then

$$\overline{A}_X = 1 - A_A.A_B - A_B.A_C - A_C.A_A + 2A_A.A_B.A_C$$

With complicated systems, as in Figure 12.11, a truth table can be drawn. For the truth table in Figure 12.11,

$$\overline{A}_X = \overline{A}_A.\overline{A}_B + \overline{A}_A.\overline{A}_B.\overline{A}_C + \overline{A}_A.\overline{A}_B.\overline{A}_C.\overline{A}_D + \overline{A}_B.\overline{A}_C.\overline{A}_D + \overline{A}_A.\overline{A}_C$$
$$+ \overline{A}_B.\overline{A}_D + \overline{A}_A.\overline{A}_D$$

The Boolean equation may be written out directly from inspection of the diagram as follows:

$$\overline{A}_X = \{(\overline{A}_A + \overline{A}_B).(\overline{A}_C + \overline{A}_D)\} + \overline{A}_A.\overline{A}_B$$

For more complex situations consult Green and Bourne (1972).

As each component availability is usually a decimal fraction, which approaches a value of unity with improving reliability and reducing repair time, the value is often a figure such as 0.999 999 945, where only the last two figures are significant. Therefore, whether calculators or computers are used, care must be exercised to ensure that these figures are not lost because the particular hardware cannot handle the required number of decimal places, or that the significant figures are not lost by rounding.

Spreadsheet programs such as Lotus 1-2-3 or SuperCalc are very useful in producing a tabular output of failure rates, repair times, unavailabilities etc., and some examples of evaluations are given on the following pages.

If the system unavailability is multiplied by the number of hours in a year (8736), the result is the annual system downtime in hours.

12.5   Confidence limits

Failure rates are statistical values based on samples of known population. However, neither the total population, nor the mean value of failure rate for all components of a particular type, nor the way the values vary over the range from the worst to the least, is known. The electrical engineer needs to know how closely the sample mean $\bar{x}$ agrees with the total population mean value of failure rate μ. Failures in similar components will tend to reach a peak after a certain time and then tail off again, producing a bell shaped characteristic called a *normal distribution* (see Figure 12.8).

Various statistics may be calculated from the data available. Those of particular interest here are as follows. First, the sample variance is given by

$$s^2 = \frac{(x_1-\bar{x})^2+(x_2-\bar{x})^2+\ldots+(x_n-\bar{x})^2}{n-1}$$

The sample standard deviation s is obtained by taking the square root of the variance. The true population variance is usually denoted by σ.

Now 95% of the standard normal distribution lies between -1.96 and $+1.96$, so the interval between $x - 1.96\sigma/\sqrt{n}$ and $x + 1.96\sigma/\sqrt{n}$ is called the 95% *confidence interval* for μ. Given the sample mean $\bar{x}$, this means

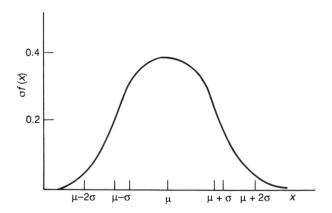

Figure 12.8 Normal distribution characteristic

that we are 95% confident that this interval will contain μ. The two end points of the confidence interval are called the confidence limits. In practice, failure rates are only known for samples, so the standard deviation is unknown and the sample standard deviation is used to estimate σ. Although the 95% confidence interval briefly discussed above is for cases where σ is known, the interval when σ is unknown approaches the same value as the sample size increases, as shown in Table 12.1.

There is, however, one more item to take care of before the confidence limits can be established. Each new sample taken will yield a new set of results with different values for the sample variance each time. It can be shown that this statistic follows a distribution called the chi-squared distribution (see Figure 12.9). This distribution is related to the normal distribution and depends on a parameter known as the *degrees of freedom*. Degrees of freedom is usually abbreviated to DF, and we would say that the estimate s^2 of σ^2 has $(n-1)$ DF.

From statistical tables, it is possible to attach confidence limits to failure rates as in the following example.

Example 12.1
A sample of 12 equipment failure rates is provided in a databank. The mean failure rate $\bar{x}$ value given is 70 per 10^6 hours. The sample standard deviation s is $0.56/10^6$ hours. Calculate the 95% confidence limits.

Solution
From Table 12.1 for a sample of size 12, the 95% confidence interval for the true population mean μ is given by $\bar{x} \pm 2.2(s/\sqrt{n})$ or 70 ± 2.2 (0.56/3.464), which gives 70 ± 0.356 failures per million hours.

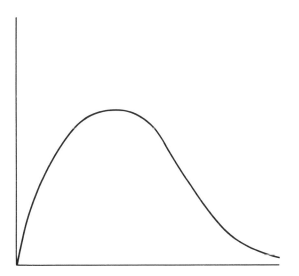

Figure 12.9 Chi-squared distribution characteristic

**Table 12.1 Values of the percentage point of distribution t_c
for a 95% confidence interval**

n	t_c
4	3.18
8	2.36
12	2.2
20	2.09
∞	1.96

Example 12.2

A large chemical plant has four main areas of activity which are grouped geographically. It is important that interruptions in supply to any of the four areas are kept to a minimum and that those that do occur are short-lived. Assuming standby generation is required, which of the two configurations in Figure 12.12 gives the better reliability, assuming all other considerations have equal effect? Note that each arrangement has the same number of cables, generators and transformers.

Solution

With the simple example, where component availabilities are identified and the availability effects due to configuration can easily be seen, Figure 12.12(a) may be identified by inspection as the more reliable, since at least two components must fail before any part of the system is blacked out. Nevertheless, in order to illustrate the methods of calculation, the two arrangements will be analysed.

The configurations in Figure 12.12(a) and (b) may be converted into reliability block diagrams as in Figure 12.13(a) and (b) respectively. As indicated in Section 12.4, a reliability block diagram represents the configuration in terms of Boolean logic. Components may be repeated so as to represent a system success (i.e. 'supplies available at all busbars') path through the diagram from left to right. As with all Boolean diagrams, the integrity of the system may be 'tested' by deleting components to see if a path or paths remain through the network.

The level of detail used in the analysis should be related to the reliability data available. For example, circuit breakers may be introduced as components which may appear on the reliability diagram in different locations depending on the mode of failure. If equipment is located in the

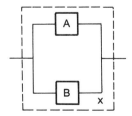

Figure 12.10 Reliability block diagram for two components in parallel

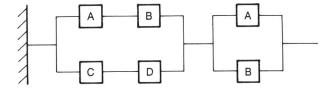

Figure 12.11 For more complex arrangements a truth table may be used:

A	B	C	D	System success	
1	1	1	1	1	
×	1	1	1	1	
×	×	1	1	×	$\overline{A}_A \cdot \overline{A}_B$
×	×	×	1	×	$\overline{A}_A \cdot \overline{A}_B \cdot \overline{A}_C$
×	×	×	×	×	$\overline{A}_A \cdot \overline{A}_B \cdot \overline{A}_C \cdot \overline{A}_D$
1	×	×	×	×	$\overline{A}_B \cdot \overline{A}_C \cdot \overline{A}_D$
1	1	×	×	1	
1	1	1	×	1	
×	1	×	1	×	$\overline{A}_A \cdot \overline{A}_C$
1	×	1	×	×	$\overline{A}_B \cdot \overline{A}_D$
1	×	×	1	×	$\overline{A}_B \cdot \overline{A}_C$
×	1	1	×	×	$\overline{A}_A \cdot \overline{A}_D$

same substation or cables in the same trench etc., blocks may be introduced to represent common mode failures such as substation fires or cable excavation accidents.

Evaluation of alternative configurations
The following numerical values are taken for this example:

Component	$\mu = 1/MTTR$	$\lambda (\times 10^{-6})$	A	$\overline{A}$
Generator	0.04166	50	0.998801	0.001199
Cable	0.1666	10	0.999940	0.00006
Busbar	0.1666	0.175	0.9996389	0.000361
Transformer	0.01	0.13	0.999998	0.000002

Using these illustrative values, we can set out the data for Figure 12.11(a) as in Table 12.2. The total unavailabilities are summarized in Figure 12.14(a). The final equation is as follows, where the numbers 1–5 now refer to Figure 12.14(a):

$$\overline{A}_{(a)} = \{(\overline{A}_1 + \overline{A}_2) + (\overline{A}_1 + \overline{A}_5 + \overline{A}_4)\} \cdot \{(\overline{A}_3 + \overline{A}_4) + (\overline{A}_3 + \overline{A}_5 + \overline{A}_2)\}$$
$$= \{2\overline{A}_1 + \overline{A}_2 + \overline{A}_4 + \overline{A}_5\} \cdot \{2\overline{A}_3 + \overline{A}_2 + \overline{A}_4 + \overline{A}_5\}$$
$$= \{0.003312\} \cdot \{0.005706\}$$
$$= 1.8898272 \times 10^{-5}$$

This figure may be expressed as downtime in hours per year by multiplying by 8760, to give 0.1655 hours or 10 minutes per year.

The configuration in Figure 12.13(b) may be similarly summarized as in Figure 12.14(b), and in this case the final equation is

$$\overline{A}_{(b)} = \overline{A}_1.\overline{A}_3 + \overline{A}_2 = 1.62407 \times 10^{-3}$$

The downtime is therefore 14.18 hours per year.

Hence the configuration of Figure 12.13(a) is 14.18/0.1445 = 98 times more reliable than that of Figure 12.13(b).

Table 12.2 Evaluation of Figure 12.13(a)

Component	$\mu = 1/MTTR$	$\lambda\ (\times 10^{-6})$	A	$\overline{A}$
Transformer	0.01	0.13	0.999 998	0.000 002
Cable 1	0.166 6	10	0.999 940	0.000 06
Bus A	0.166 6	0.175	0.999 638 9	0.000 361
Cable 3	0.166 6	10	0.999 940	0.000 06
Bus C	0.166 6	0.175	0.999 638 9	0.000 361
Total (LHS top)	–	–	–	0.000 844
Generator	0.041 66	50	0.998 801	0.001 199
Cable 2	0.166 6	10	0.999 940	0.000 06
Bus B	0.166 6	0.175	0.999 638 9	0.000 361
Cable 4	0.166 6	10	0.999 940	0.000 06
Bus D	0.166 6	0.175	0.999 638 9	0.000 361
Total (LHS bottom)	–	–	–	0.002 041
Bus C	0.166 6	0.175	0.999 638 9	0.000 361
Cable 3	0.166 6	10	0.999 940	0.000 06
Bus A	0.166 6	0.175	0.999 638 9	0.000 361
Total (RHS top)	–	–	–	0.000 782
Bus D	0.166 6	0.175	0.999 638 9	0.000 361
Cable 4	0.166 6	10	0.999 940	0.000 06
Bus B	0.166 6	0.175	0.999 638 9	0.000 361
Total (RHS bottom)	–	–	–	0.000 782

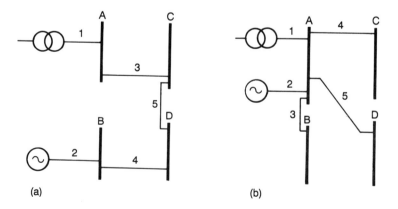

(a) (b)

Figure 12.12 Single line diagrams for Example 12.2.

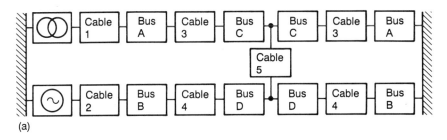

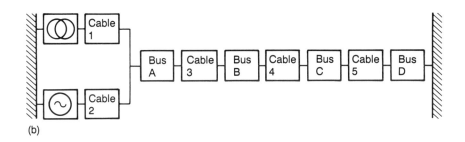

Figure 12.13 Reliability block diagrams for Example 12.2

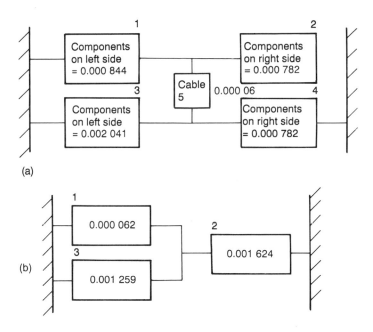

Figure 12.14 Numerical evaluation of Example 12.2

Chapter 13

Commissioning

This chapter is intended as a general checklist for use when preparing commissioning workscopes, and should not be treated as exhaustive. Every piece of equipment, and the way that it is intended to be operated, will need to be thoroughly understood before writing such a document. Any document of this kind should also include reference to access and permit-to-work procedures specific to the installation where the commissioning is being carried out.

All test equipment used should have valid calibration certificates, since, apart from the results of tests being invalid, incorrect equipment control or protective device settings could render the equipment unsafe and in some cases create a fire or explosion hazard. The test equipment itself, when used in areas where flammable gas may be present, will also be an ignition hazard, and therefore 'hot work' permits will be required for most of the tests described.

It is important that all test results are recorded so that a complete maintenance history can be retained for reference, should a failure occur. Some typical test sheets are shown in Appendix B. Lastly, all commissioning must accord with the client oil company's standard procedures, the relevant British Standards and the manufacturer's installation and operating manuals for the equipment concerned.

13.1 Generators

At the time of testing, it is assumed that the manufacturer has already successfully conducted the following tests:

(a) open-circuit test;
(b) iron loss test;
(c) short-circuit test;
(d) transient performance tests (sudden loading and three-phase short-circuit);
(e) voltage balance;
(f) heat run at full power;
(g) waveform analysis;
(h) vibration analysis;

(i) noise test;
(j) friction and windage losses;
(k) AVR and control panel testing and calibration.

It may be necessary to repeat some of these tests in the module yard and/or offshore if the machine is damaged or tampered with during transportation or installation.

Having finally reached the offshore commissioning stage, it will be necessary to make the generator safe for testing by carrying out the following:

1. Assuming the generator has already been taken out of service, the generator circuit breaker should be locked in the earth position and the associated automatic voltage regulator voltage transformers (VTs) should be isolated and padlocked. If the circuit breaker is not of the truck type it must be padlocked off and a separate earth safely applied. VT fuses should be removed, if the VT is not withdrawable.

2. It must be remembered that a large offshore generator will develop an appreciable voltage from its residual field if it is allowed to rotate. This is less unlikely than it sounds, especially if, at the same time, another commissioning team is anxiously attempting to complete the commissioning of the prime mover. Although barring gear is normally provided on the larger machines and may be used to prevent rotation, it may be necessary to provide some other temporary means of preventing rotation. The start permissive key should be removed from the prime mover control panel. All such precautions should be included in the permit-to-work system in use on the installation.

3. When it is certain that everything is electrically and mechanically safe, work can then begin on testing the generator windings.

4. Cold resistance of main stator windings. The method involves injecting a known DC current through each winding and measuring the voltage drop through the winding at the terminals. Before applying the current, the earth must be removed from the winding. Readings of voltage and current should be recorded for all three phases. The three resistances calculated from these readings, after compensating for ambient temperature, should not deviate from each other by more than 2% or from the manufacturer's design figure by more than 5%. For compensation purposes, the ambient air temperature will need to be measured at shaft height and one metre from the alternator frame.

5. Rotor and exciter field windings. The resistance of these windings should be measured in the same way as the stator windings.

6. Using an AC high-voltage test unit, the generator windings should be tested to twice their working voltage plus 1000 volts, with a minimum of 1500 volts as required in BS 4999. The windings should withstand this voltage for 1 minute. The voltage should be gradually applied, and the minute count started once full test voltage has been reached.

7. The exciter field should be tested in the same way and using the same criteria after temporarily removing the AVR connections.

8. With a brushless machine, the rotor and exciter armature should be checked for continuity only, in order to avoid damage to the diode bridge.

9. Anti-condensation heaters should be high-voltage tested to about 1500 volts to the machine frame, with adjacent windings earthed. A continuity check should also be carried out.
10. Using a megger of suitable voltage rating (5000 V for medium-voltage and 1000 V for low-voltage machines), the insulation resistance should be measured. Minimum values acceptable are as follows: stators 50 MΩ; exciter field 10 MΩ; exciter armature 10 MΩ; rotor 10 MΩ.

Note that during any of the high-voltage or megger tests, all voltage sensitive auxiliaries, thermistors, auxiliary windings etc. should be temporarily earthed to avoid damaging them. No part of the AVR or rotating diode bridge should be meggered, if damage is to be avoided.

A check must be made that all non-electrical parts of the generation system are complete and functioning as follows:

(a) fuel system;
(b) engine controls including the governor and overspeed protection devices;
(c) engine and alternator cooling system;
(d) ventilation system;
(e) air intake and exhaust systems;
(f) engine starting equipment;
(g) generator package fire and gas detection and protection systems.

Assuming the associated switchgear has already been commissioned, the phase and earth cabling and conductors, including neutral earthing resistors, should be checked for continuity and insulation.

Before the prime mover is started, the generator circuit breaker should be racked into the test position, the associated voltage transformers racked in, fuses replaced, and the means of preventing generator rotation removed.

With the prime mover running, the following equipment may be commissioned:

1. Under/overvoltage protection check. (Note that this should have already been checked beforehand by primary injection.)
2. Automatic synchronizing unit. (Note that with the circuit breaker in the test position, this consists of observing that the circuit breaker closes at the correct phase angle and slip frequency limits.) A second generator is required for a reference.
3. Manual synchronizing equipment and check synch relay.
4. Excitation system, including AVR setting.
5. Over/underfrequency protection check.
6. General check that control panel instrument readings are normal for the generator on no-load.

For generator load testing, it is advisable to connect a suitable load bank to the associated switchboard (or direct to the alternator if the switchboard is already in use), so that the electrical and mechanical stability of the package may be checked before it supplies load to platform consumers. The load bank must include both resistive and reactive switchable elements.

13.2   Switchgear

13.2.1 Air break switching devices and fuses rated for use at voltages below 1000 V

First, the equipment should be given a thorough mechanical and visual inspection. This should include the following:

1. The equipment should be clean, dry and free from extraneous loose items or tools.
2. Insulators and insulation should be clean and dry.
3. Indicating, protective and interlocking devices should operate satisfactorily.
4. All conducting connections should be secure and intact.
5. The ratings of fuses, relays and other devices should be as shown on design drawings.
6. Cables should be of adequate cross-section and correctly glanded and terminated.
7. Internal wiring should be correctly secured.
8. Switching contacts should normally be sparingly applied with a suitable contact lubricant, but the manufacturer's manual (or his advice) should be consulted, since in some conditions this may lead to contact sticking or reduced fault clearing capability.
9. The mechanical components of switching and interlocking mechanisms should function smoothly and correctly.

Electrical tests will then need to be completed as follows:

1. A primary insulation test should be carried out on each phase to ground and from each phase to phase using the test voltages shown in Table 13.1.

Table 13.1 Recommended insulation resistance test voltages (BS 5405:1976)

Three-phase voltage rating of primary insulation (kV)	*Test voltage recommended for insulation resistance test to earth and between phases* (kV DC, 1 minute)
Up to 1	1
Above 1 and up to 3.6	2
Above 3.6 and up to 12	5
Above 12	10

2. The insulation resistance of small wiring and ancillary components should be tested using a megger or similar device, where the test voltage is not greater than 500 V DC.
3. The overloads of motor control gear should be injection tested at 100%, 125% and 600% setting values, and checked against the manufacturer's tripping times. Each thermal or magnetic element should be tested individually. Tests made on thermal elements should include for both 'hot' and 'cold' operating conditions.
4. The switchboard metering, i.e. all ammeters, voltmeters, CTs, VTs etc., should be checked for correct calibration.

5.   A DC overvoltage test of 2000 V should be applied for 1 minute to all load carrying busbars and circuit breakers. The test should be applied between phases with the circuit breaker closed, and both between phases and from phases to earth with the circuit breaker open. Any auxiliary equipment which inherently cannot withstand the 2000 V DC test voltage (e.g. semiconductor devices) should be disconnected before the start of the test. Voltage transformers should be isolated by removing both primary and secondary fuses. Current transformer secondaries should be fitted with short-circuiting links, and earthed before starting the test.

6.   When the electrical testing is complete, each protective interlocking and tripping device such as overcurrent, earth fault and undervoltage releases should be operated to its full extent in order to prove its function before the switchboard is brought into service.

13.2.2 Air circuit breakers and switchgear rated for use at voltages above 1000 V

The equipment should first be inspected visually and for correct mechanical operation as follows:

1.   Inspect for physical damage and compare nameplate details with platform design document requirements.
2.   Make sure that the switchgear cubicle has been correctly installed and adequately anchored to the switchroom floor to withstand shock loadings imposed by operation of circuit breakers.
3.   Referring to the manufacturer's manuals, perform all the mechanical operator and contact alignment tests on both the circuit breaker and its operating and interlocking mechanisms, in accordance with the manufacturer's instructions.
4.   Check that the operation of all safety shutters is satisfactory by removing circuit breaker trucks and physically checking that shutters are free to move easily.
5.   Check that all insulation and insulators are clean and dry.
6.   Check that main contacts, secondary contacts, auxiliary switches and earthing contacts are correctly fitted and aligned.

Electrical tests will need to be carried out as follows:

1.   Measure the resistance between pairs of contacts. This should be within the range given in the manufacturer's manual, and not more than 500 $\mu\Omega$ if excessive contact heating is to be avoided.
2.   Perform a voltage pickup test on the tripping and closing coils.
3.   Trip the circuit breaker a number of times by operating each protective device in turn.
4.   Measure the primary insulation resistance for pole to ground, pole to pole, and across open poles of the same phase.
5.   Perform insulation resistance tests at 500 V DC (using a megger or similar instrument) on all control wiring. Circuits containing semiconductor devices such as solid state protection relays should not be tested. These tests should be performed using the voltage values given

in Table 13.1, resulting in a minimum value of insulation resistance of 100 MΩ. The above test should be performed both before and after a high-voltage DC test. The first application will determine if the insulation resistance is high enough for acceptance, and whether the high-voltage test may be performed. The second test will verify that the application of the overpotential test voltage has not shown up any weakness in the insulation. In each case, the test voltage should be applied for 1 minute.

6. Overpotential test voltages as shown in Table 13.2 should be applied to phase conductors with circuit breakers in the open and closed positions. Arc chutes should be tested for watts loss within the manufacturer's allowable values. The test voltages in Table 13.2 apply to metal enclosed switchgear and control gear (i.e. assemblies with external metal enclosures intended to be earthed and complete except for external connections). The switchgear under test should be free from all external cabling, voltage transformers, current transformers and other auxiliary equipment which cannot withstand the test voltage being applied.

Table 13.2 Recommended DC test voltages on metal enclosed switchgear and control gear after erection on site (BS 5227:1975 and BS 5405:1976)

Rated voltage U (kV RMS)	Site test voltage (kV DC, 15 minute)
Up to 1	1
Above 1 and up to 3.6	2
Above 3.6 and up to 7.2	7.5
Above 7.2 and up to 12	15
Above 12 and up to 17.5	25
Above 17.5 and up to 24	32
Above 24 and up to 36	45
Above 36	66

If main cabling terminated at the switchgear is to be high-voltage tested, then those parts of the equipment which cannot be readily isolated from the main cable terminals should be capable of withstanding the DC test voltage specified in BS 5227 for 15 minutes (i.e. the duration of the test).

When any overpotential testing is undertaken, the test voltage should be increased as rapidly as consistent with its value being indicated by the test instrument. If the voltage breaks down the insulation as it is being increased, the test set operator should have raised the voltage slowly enough to have been able to record the value at which the breakdown occurred.

Table 13.2 gives the maximum DC site test voltage, and this value should be maintained for 15 minutes. Test current values should be recorded at 30 second intervals for the first 2 minutes and at 1 minute intervals thereafter. The test potential should then be reduced to zero and earths applied for 10 minutes before continuing the next test.

Although it is usually obvious if the equipment fails this test, values of leakage current equivalent to an insulation resistance greater than 100 MΩ are expected, and leakage current at the full test voltage should remain steady. If the leakage current gradually increases during the test, this is likely to indicate an incipient failure or the connection of a semiconductor or similar component which should have been isolated before starting the test.

13.2.3 Other switchgear

Commissioning checks for other forms of switchgear follow a similar pattern, but with the following exceptions.

13.2.3.1 Oil switchgear over 1 kV

The mechanical check should include a visual check of the circuit breaker tank oil levels and an examination of oil filled bushings for any leakage.

The electrical testing should include an oil sample insulation test in accordance with BS 5730:1979.

13.2.3.2 Vacuum circuit breakers

The mechanical check should include a visual inspection of the interrupters for damage or seal corrosion. If fitted, the wear gauge should be checked in each case.

The electrical testing should include vacuum integrity tests by application of a DC overpotential (usually around 20 kV). The manufacturer should be consulted as to both the correct voltage to apply and the safety precautions that should be taken owing to the possibility of generating X-rays during this test.

13.2.3.3 Sulphur hexafluoride (SF₆) switchgear

The mechanical inspection should include the following:

1.   Check that the gas system is operating at the correct pressure.
2.   If the switchboard is of the sealed type, topping up of SF_6 should be required infrequently by using a temporary connection to a gas cylinder. If the gas system has a topping up compressor, the compressor gas supply, oil level, filters or desiccants etc. should be checked for good condition. The permanently fitted gas pressure gauge should be checked against a recently calibrated test gauge.

As contact mechanisms tend to be inaccessible in this type of switchgear, electrical testing should include timing and travel tests, subject to the unit being properly equipped for this to be carried out. The results may then be compared with those provided by the manufacturer, discrepancies indicating a fault in operating mechanism or contact wear.

13.3   Protection schemes: general procedures

The satisfactory operation of protection relays is important for the prevention of danger to equipment operators and local damage to equipment. By preventing an electrical fire, it may also have prevented growth of an incident to platform threatening proportions. Therefore great care should be exercised by the commissioning engineer to ensure that the protection devices will respond to electrical system disturbances as the designers intended. If the commissioning engineer considers that he has found design errors or omissions during commissioning, they must not be shrugged off but must be discussed with the designers and/or operators.

The testing of voltage transformers, current transformers and protection relays is described in detail in the following sections.

The general procedure for the commissioning of protection schemes is to perform the following checks and tests:

1. Obtain a full set of wiring diagrams, schematics, and the protection relay setting schedules for the equipment to be tested.
2. Ensure that you are in possession of the manufacturer's technical data and commissioning procedures.
3. Carefully inspect the overall scheme, checking all connections and wires on CTs, VTs and protection relays.
4. Measure the insulation resistance of all circuits.
5. Carry out ratio, polarity and magnetization curve tests on the current transformers.
6. Carefully inspect each relay before testing it by secondary injection in accordance with the manufacturer's installation instructions.
7. Check the stability of the protective relays for external faults and determine the effective current setting for internal faults, by means of primary injection tests.
8. Perform closing, tripping and alarm checks for the equipment associated with protective relays.

Before commencing work on any equipment, the isolation and permit-to-work procedures must be strictly adhered to. In particular, current and voltage transformers should be de-energized, isolated and discharged to earth before any commissioning operations are carried out.

13.4 Voltage transformers

First the VT should be inspected both visually and mechanically as follows:

1. Inspect for physical damage and for compliance with design documents.
2. Check mechanical alignment, clearances and proper operation of any disconnecting and earthing devices associated with the VT.
3. Verify proper operation of grounding or shorting devices.

After the initial examination has been carried out, the following electrical tests should be completed.

13.4.1 Insulation resistance

1. Remove all deliberate earth connections from the VT windings.
2. Electrically isolate the VT primary and secondary windings by removing fuses, links or connections.
3. Perform insulation resistance tests on secondary windings to earth and between secondary windings at 500 V DC for 1 minute.

4.   Having obtained approval from the VT manufacturer, perform a high-voltage DC insulation test on the primary winding insulation. Recommended test voltages are given in Table 13.3, which is for fully insulated windings. This test is only to be used on new VTs which have yet to be placed in service, and is for 15 minutes duration. Secondary windings should be shorted together and to earth for the high-voltage DC test, so that the insulation is tested with a potential from primary to earth and primary to (earthed) secondary. The test voltage should be gradually increased up to the full value and maintained for the duration of the test.
5.   An insulation test of 500 V DC should be applied to all windings to earth, and between windings, for a duration of 1 minute. This test should be performed both before and after the high-voltage DC test described above, and should result in a minimum resistance reading of 100 MΩ.

Table 13.3 Recommended test voltages for voltage transformers (fully insulated primary) and current transformers

Rated highest equipment voltage (kV AC)	Test voltage (kV DC)
1. Primary insulation test	
Up to 0.66	1.5
Up to 3.6	7.5
Up to 7.2	15.0
Up to 12.0	25.0
Up to 17.5	32.0
Up to 24.0	45.0
Up to 36.0	66.0

1. The application time should be 15 minutes applied between the primary and earth, with secondary windings shorted and earthed.

2. Secondary insulation tests. All secondary insulation tests should be conducted using a test voltage of 500 V DC to earth for one minute duration

13.4.2 Polarity check

The VT polarity can be checked using the simple circuit shown in Figure 13.1.

The battery must be connected to the primary winding with the polarity ammeter connected to the secondary winding. For the circuit shown, the ammeter will flick positive on making the switch, and negative on opening the switch.

13.4.3 Voltage ratio test

The approximate voltage ratio of the transformer can be tested by a simple voltage injection test on the primary winding, by measuring the primary and secondary voltages.

Extreme care must be exercised when carrying out this test to ensure that the test voltage is applied to the primary and *not* the secondary

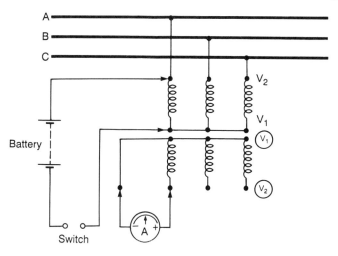

Figure 13.1 Circuit diagram for carrying out VT polarity checks: for polarity shown, ammeter flicks positive on switch make and negative on switch break

winding. Otherwise, very high voltages may be applied to the test instruments and create a hazard for the tester.

13.4.4 Phasing check

It is essential to check the phasing of the secondary connections on three-phase VTs, or three single-phase VTs, when these are used for either metering or protection purposes.

The simplest method of testing the phasing is by the application of the system voltage on to the primary winding of the transformer, i.e. with the main busbars live. The secondary voltages can then be checked between phases and neutral, and the phase relationship verified with a phase indication meter.

Open-delta secondary windings of three-phase transformers can also be checked to verify a balanced primary three-phase supply.

13.4.5 Restoration

Restore all electrical connections to their correct state after having removed any temporary test wiring.

Check all fuses and current limiting resistances, if provided, for continuity, correct rating and good condition. Replace any component which shows signs of thermal or mechanical damage.

13.5 Current transformers

Visual, mechanical, insulation resistance and polarity checking are carried out in a similar way to the methods for the voltage transformers discussed above. Figure 13.2 shows the connection arrangements for polarity checking.

The following tests should also be undertaken.

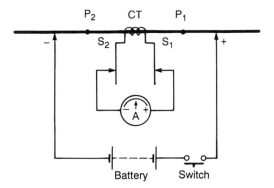

Figure 13.2 Circuit diagram for carrying out CT polarity checks: for polarity shown, ammeter flicks positive on switch make and negative on switch break

13.5.1 Current ratio check

Two methods may be adopted as follows.

13.5.1.1 Method 1
This test is performed by secondary injection, as shown in Figure 13.3. Current is passed through two primary windings using a primary injection test set, and the primary and secondary currents are measured. A temporary short-circuit is placed across the primary windings at one end and the injection is applied at the other. The ratio of primary to secondary

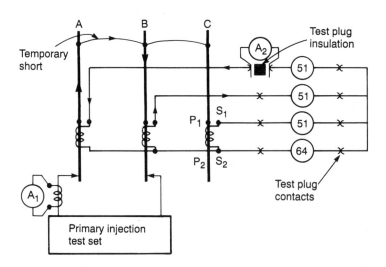

Figure 13.3 Circuit for current ratio check on CT by primary injection (method 1)

ammeter readings should be approximately equal to the marked ratio on the current transformer nameplate.

This method of current ratio test also provides a second check on polarity, since the residually connected secondary ammeter will read a few milliamperes if the current transformers are of the correct polarity, but a value approximating to the CT ratio current if the polarity is incorrect. In some cases, such as in metal clad switchgear, it may not be possible to access the transformers physically, in which case the protection circuit drawings should confirm that the polarity is correct.

13.5.1.2 *Method 2*
An alternative method is to perform a primary injection test on one CT only, as shown in Figure 13.4. As with method 1, the ratio of primary to secondary test currents should approximate to the transformer ratio marked on the nameplate. For multitapped CTs, the ratio should be checked at each tap.

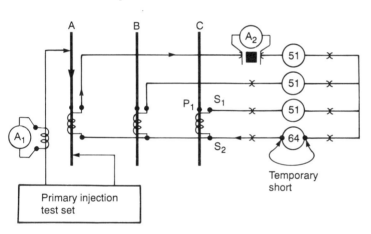

Figure 13.4 Circuit for current ratio check on CT by primary injection (method 2)

13.5.2 Magnetization curve tests

The current transformer magnetization curve should be checked at several points so as to establish the approximate position of the knee point of the transformer.

The test circuit for obtaining this point is shown in Figure 13.5. A test voltage is applied to the CT secondary winding, and the resulting magnetizing current is measured with the primary winding open-circuit. The applied voltage is increased gradually until the magnetizing current is seen to increase rapidly for small increases in voltage, indicating that the knee point has been reached. The test voltage should then be reduced carefully in stages. The voltage and current should be recorded at several points both above and below the knee point and down to zero test voltage.

The knee point is defined as the point on the CT magnetization curve where a 10% increase in applied voltage results in a 50% increase in

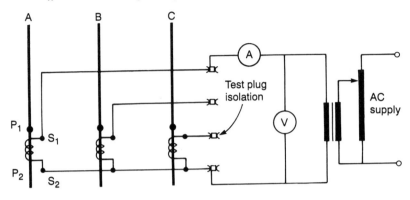

Figure 13.5 Test circuit for obtaining a CT magnetization curve

magnetizing current. Generally, the knee point level should be reached when the secondary voltage is raised until the magnetizing current is equal to or less than the rated secondary current.

Typically, the magnetizing current is about one-tenth of the secondary current rating. The test voltage required may be higher than 240 V, in which case a step-up transformer will need to be included in the testing kit.

After the magnetization curve check has been carried out, the secondary winding resistance of the CT should be measured.

13.5.3 Restoration

When all the tests have been completed satisfactorily, the test connections should be removed and all permanent electrical connections should be restored to their correct state.

13.6 Protection relays

In general, secondary injection is used to check the operation of relays via test blocks. An overcurrent relay, for example, can be checked against the manufacturer's operating curve and tolerances for various current taps and time multiplier settings.

Once the relay is operating correctly by secondary injection, primary injection tests may then be performed in order to determine the effective settings for internal faults, whilst proving the stability of the protection for external faults.

The most common element types used in protection relays are:

(a) the induction disc;
(b) the attracted armature;
(c) the bimetallic thermal element;
(d) electronic circuitry which replicates the characteristics of the above electromechanical devices.

One relay case may contain a combination of different elements to meet the particular protection requirements of the equipment it serves, and therefore a number of different test procedures may be involved in the testing of one relay case. The following sections provide only the general principles on relay testing, and manufacturers' literature and the references given in the Bibliography for Chapter 9 should be studied for a deeper appreciation of the subject. A guide to the principal applications of protection relays offshore is given in Chapter 9.

13.6.1 Mechanical checks on induction disc and attracted armature relays

1. Check that all fixing and terminal screws are secure.
2. Check that air gaps are free of dirt and foreign matter.
3. Check that the hair spring on induction discs is reasonably spiral in shape, with no turns touching.
4. Check that turn multiplier dials have a smooth action and show no sign of sticking.
5. Ensure that all the transit packing has been removed.
6. On attracted armature relays check that, when the armature is closed with a finger, normally open contacts close and normally closed contacts are opened properly by their push rods.
7. Ensure that any trip indicator flags and their latching mechanisms operate satisfactorily, and that they may be unlatched by pushing the reset rod.
8. With the aid of the system schematic drawing, check that the relay wiring is as intended.

13.6.2 Electrical checks on electronic (static) and induction disc relays

13.6.2.1 Insulation tests
A cautious approach is required when applying insulation test voltages to protective relays, especially when they contain semiconductor devices.

Secondary wiring insulation can usually be tested at 500 V DC, however, and tests should be made on relay coils, contacts and between AC and DC circuits with secondary CT and VT connections removed. The resistance values obtained should be in excess of 10 MΩ.

13.6.2.2 Secondary injection tests
These tests are normally carried out using test plugs. However, for relays which cannot be withdrawn from their cases, the test connections must be made directly on to the relay terminals.

If secondary injection tests are to be carried out on relays whose CTs are energized, then care must be taken to ensure that the CT secondary connections on the test plug are shorted out by links, so as to prevent high voltages appearing across the CT secondary when the test plug is inserted.

Overcurrent induction disc relays used offshore usually have a current/time operating characteristic in accordance with BS 142:1983 (Section 3.2) for an inverse definite minimum time (IDMT) relay. There are, however, relays with other inverse time characteristics in use, and so it is important that the manufacturer's commissioning and operating documentation is immediately at hand during any testing.

The arrangements for performing tests on the various elements are as follows. Figures 13.6 and 13.7 show how the test circuit is connected.

1. With the time multiplier (TM) at 1.0 and 100% of CT secondary current, apply 2 times current setting. The operating time should be checked and compared with the relay characteristic.
2. With the TM at 1.0 and 10 times current setting, check the operating time against the characteristic once again. In each case, the relay operating time should be within the allowable tolerance band of −7.5% to +7.5%.
3. If the relay is within tolerance, the characteristic may be checked over a range of values so that the actual characteristic may be checked against the manufacturer's standard.
4. The minimum operating current of the relay should be checked to ensure that it does not exceed the maximum tolerance given by the manufacturer.
5. Check that the maximum current at which the relay will reset immediately following operation (the resetting current) is as stated in the manufacturer's data.
6. Check that the operating time required by the system design, shown in the relay setting schedule, is obtainable, by inspecting the measured characteristic.

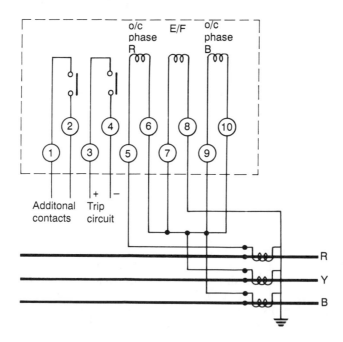

Figure 13.6 Typical triple-pole overcurrent and residually connected earth fault relay

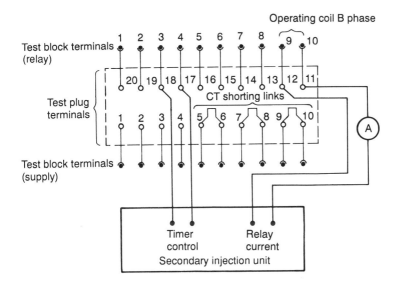

Figure 13.7 Secondary injection test circuit for triple-pole overcurrent and residually connected earth fault relay

7. If there are any attracted armature and/or electronic instantaneous elements in the relay, these should be tested for pickup (i.e. operation at the minimum specified voltage or current). Attracted armature relays used for this purpose are designed so as not to chatter, and if chatter occurs it is likely that the relay contacts need to be cleaned.
8. Auxiliary relays for operation of flag indicators, remote indication circuits etc. may require a separate supply in order to be tested.

13.6.3 Electrical tests on attracted armature relays

Attracted armature relays are used as lockout, tripping, control, counting and indicating relays, designed to sense variations in either system current or voltage. Such a relay may be located either as a single unit in its own case, or as one of a number of relays within a switchboard cubicle.

13.6.3.1 Insulation testing
With all secondary wiring disconnected from the relay, a 500 V DC insulation test should be carried out on all terminals of the relay unit. The resistance value to earth should exceed 10 MΩ.

13.6.3.2 Secondary injection testing
In general, voltage operated relays should be tested to ensure that they operate satisfactorily at 75% to 120% of normal coil rated voltages for DC operation, and 80% to 115% for AC operation. Where voltage measuring

relays are used, the pickup and dropout voltages should be checked against calibrated values. (The dropout voltage is the voltage at which the relay coil is no longer able to hold the armature in its energized position, so that it opens or drops out.)

Current operated relays should be tested to ensure that minimum operating current does not exceed the manufacturer's recommended value. In the case of auxiliary relays, tapped coils are provided and the pickup current should not exceed the value recommended for the particular tap. For current measuring relays, the pickup/dropout ratio should be checked against the manufacturer's reference value.

13.6.4 Electrical tests on thermal relays

Thermal relays are frequently used for motor overload protection, and may be designed to provide protective sensing for single phasing, stalling and unbalanced loading. The P&B Golds relay is the most common electromechanical type, and the testing information given below reflects this. However, electronic relays, often with some data processing capability, are now available in production quantities, and these are designed to replicate the thermal characteristic of the motor in a similar manner.

The following tests would usually be required as a minimum:

1. Check that each heater element operates in accordance with the calibration curve, from cold, by applying a multiple of the rated current through the elements, having connected them all in series. The time to indicate a given percentage (say 115%) of rated relay current whilst the above multiple of the rated current is applied, should be within the range quoted in the manufacturer's calibration data.
2. With the heater elements still connected in series, check the hot operating time by increasing the current from 100% running current to a multiple of this representing the motor starting current (about 600% running current). Measuring the time for the relay to operate at about 115% running current will enable the hot characteristic time to be checked.
3. Check the operation of the relay at different tap positions by varying the injection current. This should confirm that the operation of each bimetal phase element is balanced with that of its neighbour.
4. Check that if the heater of each element is injected in turn, representing a single phasing situation, the relay becomes unbalanced and trips.
5. Test any attracted armature elements (instantaneous overcurrent, earth fault, flag auxiliary etc.), if fitted, for operation at the correct settings.

13.6.5 Primary injection tests

Primary injection tests are carried out on the relays following the completion of all commissioning tests on CTs and VTs and relay calibration tests by secondary injection. The purpose of the primary injection test is to check the stability of the system for external through faults and to check the entire system, including the circuit breaker control circuitry, for satisfactory operation when subjected to a fault current to which it is designed to react.

Transporting primary injection test sets offshore is often fraught with difficulties, since they tend to be too heavy to be flown out to the installation and have to be transported by sea. Having seen the treatment such equipment can be subjected to when lifted off a supply boat in heavy weather, the author would advise that great care should be taken in packing the equipment to avoid damage and hence delays to the commissioning programme.

As switchgear, cabling and protection system designs can vary greatly, design documentation and the manufacturer's manual will need to be carefully considered for each part of the system to which primary injection currents are applied.

The most common types of protection are overcurrent, earth fault (residual) and restricted earth fault, for which typical test circuits are given in Figures 13.8 to 13.10.

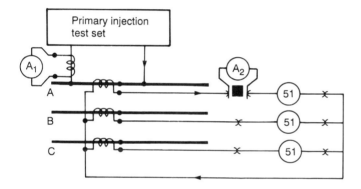

Figure 13.8 Primary injection test circuit for overcurrent relays

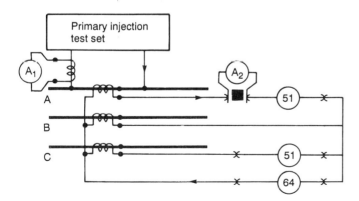

Figure 13.9 Primary injection test circuit for residually connected earth fault relays

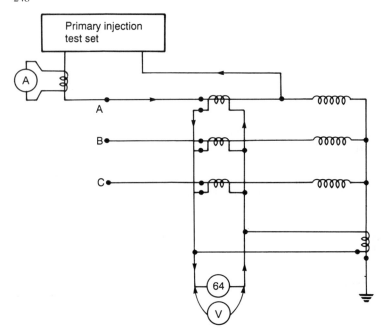

Figure 13.10(a) Primary injection test circuit for sensitivity check on a restricted earth fault relay

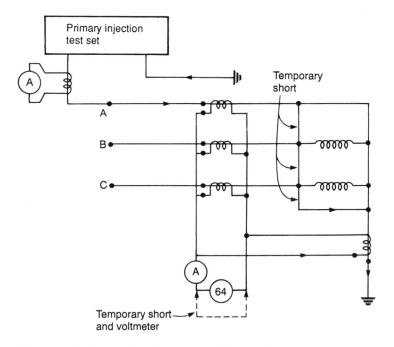

Figure 13.10(b) Primary injection test circuit for stability check on a restricted earth fault relay

13.6.5.1 *Overcurrent relays*
The effective setting for overcurrent relays can be checked using the circuit in Figure 13.8. The method shown checks the operating time of each element, and the residual current flowing in the common CT connection represents the imbalance between the phases under test. This residual current should be very low, normally less than 1 mA.

13.6.5.2 *Residual earth fault relays*
The sensitivity of residually connected earth fault relays can be tested as shown in Figure 13.9. Relays connected in this manner can be either instantaneous current operated elements or induction disc elements. The effective setting of the relay for primary faults is checked by single-phase injection.

13.6.5.3 *Restricted earth fault relays*
For restricted earth fault relays, both sensitivity and stability tests are necessary, as these relay schemes may be subject to through fault currents owing to faults in the electrical system outside their zone of operation. Their most common application offshore is on transformer windings or as part of a transformer differential protection scheme.

The circuit for sensitivity testing is given in Figure 13.10(a). Each main current transformer is injected, and the voltage across the earth fault relay and the stabilizing resistor is measured for a level of injection current which just gives relay operation.

The stability of the scheme when subject to external fault currents is tested by the application of the test circuit shown in Figure 13.10(b). Current is passed through the neutral CT and each phase in turn. With full primary load current flowing, the relay should remain stable and the ammeter should read only a few milliamperes.

13.6.6 Closing and tripping

The protection relays are fault sensing devices, and cannot interrupt fault currents themselves. Therefore it is vital that the operation of the devices which perform this duty, i.e. the circuit breakers and their tripping and closing control circuits, are checked with their associated protection relay schemes. In order that the circuit breakers can be operated, it will be necessary to commission the tripping and closing batteries and chargers prior to the primary injection test. An auxiliary DC supply is often required by the protection relays, in any case, for the operation of indication devices and as power supplies for electronic relays.

All auxiliary relays associated with each circuit breaker control system, including lockout relays and intertripping and interlocking schemes, should be functionally tested before the circuit breaker is put into service.

13.7 Motor commissioning tests

Three types of testing are recommended for the offshore commissioning of motors, as follows.

13.7.1 Megohmmeter (megger) test

This involves the DC measurement of the stator winding resistance to earth and between phases at the appropriate test voltage as shown in Table 13.4. This test is carried out on all motor windings and is of 1 minute duration. The procedure is as follows:

1. Disconnect the neutral point and adequately insulate each phase lead from earth. On small machines it will not be possible to disconnect the neutral point.
2. Disconnect and earth all resistance temperature devices (RTDs).
3. Ensure that the motor frame is connected to a known earthing point. Phases not under test must be earthed.
4. Measure the ambient air temperature in degrees Celsius local to the motor before starting the test. If the temperature of the motor is thought to be significantly higher than the motor air temperature, then the winding temperature should be measured, if possible, by using any RTDs fitted to the motor before they are disconnected for the test.
5. If the temperature of the insulation is cold to the extent that it is below the dew point, a moisture film will form on its surface and may lower the insulation resistance, especially if the winding has developed a coating of dirt and/or salt. Therefore if this is likely to occur, the motor anti-condensation heaters should be left on for some time before the motor is tested.
6. Using a megger or similar instrument, perform the following test using the voltages given in Table 13.4. With the neutral point disconnected, test the phase-to-phase insulation by applying the test voltage to each phase in turn with all other phases earthed. This test measures the combined insulation resistance of the tested phase, both to earth and to the other (earthed) phases.
7. Again using Table 13.4, with the neutral point connected, test the winding insulation to earth by applying the test voltage to each phase winding connection in turn, with all other phase connections insulated from earth. The winding insulation resistance should be corrected to 40°C by using the equation $R_c = K_t R_t$, where R_c is the insulation resistance corrected to 40°C, R_t is the insulation resistance at temperature t °C, and K_t is the insulation resistance temperature coefficient at temperature t obtained from Table 13.5.

Table 13.4 DC megohmmeter and polarization index tests for AC motors

Motor rated AC voltage (V)	DC test voltage (V)
Below 100	500
Above 100 up to 600	1000
Above 600 up to 5000	2500
Above 5000	5000

Application times:
   1 minute for insulation and *PI* tests on motors with rated voltages below 600 V;
   10 minutes for *PI* tests on motors with rated voltages above 600 V.

13.7.2 Polarization index

This test is a continuation of the megger test above, and requires the application of the appropriate test voltage for 10 minutes. The ratio of the 10 minute to the 1 minute reading is termed the polarization index *PI*, and gives some indication of the dryness of the windings and their ability to withstand overvoltage tests.

The procedure is as follows. Using a motor driven megger or similar instrument, measure the resistance to earth and between phases of the motor windings. The test voltage (obtained from Table 13.4) should be applied for 10 minutes and readings noted after 1 minute and 10 minutes.

The polarization index can be calculated as follows:

$$PI = \frac{10 \text{ minute reading}}{1 \text{ minute reading}}$$

In the case of low-voltage machines, i.e. machines whose rated voltage is 600 V or less, the *PI* is calculated using the 30 second and 1 minute readings and the application time is limited to 1 minute.

Table 13.5 Winding temperature coefficients for insulation resistance measurements at various temperatures

Winding temperature (°C)	Multiplier coefficient k_t
0	0.06
10	0.12
20	0.25
30	0.5
40	1.0
50	2.0
60	4.0
70	8.0
80	16.0
90	32.0
100	64.0

13.7.3 Overpotential tests

Overpotential tests should not be carried out on motors unless the insulation resistance is greater than 100 MΩ and the *PI* values are greater than 2.0 for class B and class F insulation, or 1.5 for class A insulation. High DC voltages are both lethal and, if the motor under test is in an area where flammable gases may be present, an ignition hazard. Carefully follow all operational procedures and safety precautions.

High-voltage DC tests should be applied as specified in Table 13.6, and the test should be made between the windings under test and the frame of the machine (which must be adequately earthed).

The test procedure is as follows. To start with, a test voltage of preferably about one-third but not more than one-half of the full test voltage quoted in Table 13.6 should be applied. This should be increased to the full test voltage as rapidly as the indicating instrument will allow without overshooting beyond the quoted value. The full test voltage must

Table 13.6 High-voltage DC tests

Machine rating	Maximum recommended DC high-voltage tests (V)
Less than 1 kVA (or kW) with rated voltage below 100 V	500 + (2 × rated voltage)
Less than 1 kVA (or kW) with rated voltage 100 V and above	1000 + (2 × rated voltage)
1 kVA (or kW) and above, but less than 10 MVA (or MW)	1000 + (2 × rated voltage)
10 MVA (or MW) and above, at rated voltages	
up to 2 kV	1000 + (2 × rated voltage)
above 2 kV and up to 6 kV	2.5 × rated voltage
above 6 kV and up to 17 kV	3000 + (2 × rated voltage)
above 17 kV	Subject to special agreement

The application of high DC voltages is subject to manufacturers' agreement. For the absolute maximum values permitted by BS 4999: Part 60, multiply the above values by 1.28.

be maintained for 1 minute, and the leakage current recorded. At the end of 1 minute, the test voltage must be reduced rapidly to not more than one third of its full value before switching off. This method of test voltage application is important in order to avoid overstressing the insulation with high transient voltages produced by switching etc.

Earths should be applied to each test winding for at least 10 minutes after completion of the test, and the complete windings for 1 hour after final completion of all tests.

BS 4999:Part 60:1976 permits DC onsite high-voltage tests to be carried out by agreement between the manufacturer and the purchaser. The test voltages must not be greater than 1.6 × 0.8 (i.e. 1.28) times the RMS value of the AC voltage specified in Table 60.11.5 of BS 4999. The values recommended in Table 13.6 are the RMS values quoted in the BS 4999 table and are therefore 80% of the maxima allowed. These values must still be agreed with the machine manufacturer, however, before testing is undertaken.

13.7.4 Evaluation of motor test results

When the tests described above have been carried out, the results must be interpreted in order to make a decision on whether or not to accept the machine for operation. An estimation of the machine's acceptability may be based on a comparison of present and previous values of insulation resistance and polarization index, corrected to 40°C in each case.

When the insulation history is not available, recommended minimum values of *PI* or of the 1 minute insulation resistance value may be used. These values are given in Table 13.7.

The insulation resistance value of one phase of a three-phase winding, with the other two phases earthed, is approximately twice that of the entire winding. Therefore, when the three phases are tested separately, the

Table 13.7

(a) Minimum insulation resistance for the entire winding at 40°C

Motor rated voltage (V)	11 000	6600	3300	440
Minimum insulation resistance (MΩ)	12	7.6	4.5	1.5

(b) Minimum polarization index values

Motor insulation class	A	B	F
Polarization index	1.5	2.0	2.0

observed resistance of each phase should be halved in order to obtain a value for comparison with the appropriate value in Table 13.7(a). (For details of guard circuits, see Thorn EMI's *A Simple Guide to Insulation and Continuity Testing*, 1984.)

The acceptance criteria in Table 13.7 must be modified, however, owing to the following limitations. First, insulation resistance is not directly related to dielectric strength, and it is therefore impossible to specify the value at which a winding will fail electrically. Secondly, the windings of large or slow-speed machines have extremely large surface areas, with

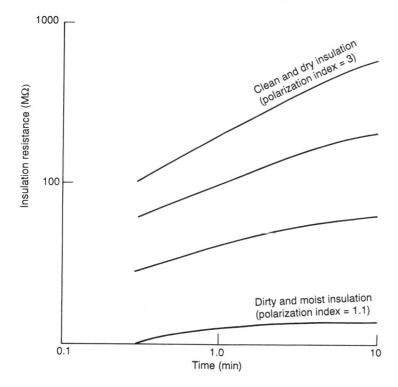

Figure 13.11 Polarization index curve for motor windings

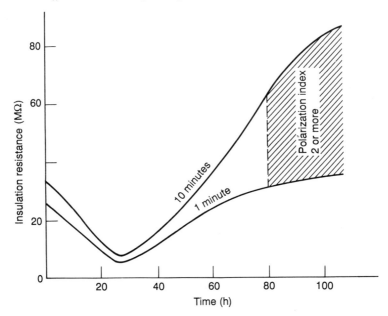

Figure 13.12 Drying curve for motor windings

healthy values of insulation resistance less than the recommended minimum. Therefore, to some extent, acceptance will depend on the individual
circumstances, such as motor size, number of poles, motor operating
environment and frequency of operation.

Typical resistance/time characteristics are shown in Figures 13.11 and
13.12, which illustrate the behaviour of insulation under different conditions, and show the significance of polarization indices. It must be stressed
that the *PI* depends upon a number of contributory factors, such as the
winding condition, the insulation class and the type of machine. Figures
13.11 and 13.12 are typical for class B insulation. Where the *PI* is low
because of dirt or moisture contamination, it can be significantly improved
by cleaning and drying. Drying curves are shown in Figure 13.12, and such
curves can be used to indicate the drying time required.

Chapter 14

Maintenance and logistics

14.1 Rationalization of spares

The weight and space limitations of an offshore installation impose restraints on the electrical equipment installed. The restraints on the stocking of consumable spares and spare parts are even more severe, however, and this must always be kept in mind at the detail design stage, particularly during the procurement of the smaller drive packages such as small pumps and compressors, distribution boards and cable accessories. Sometimes a spares rationalization can be made by using the same model of prime mover for main generators and gas lift or export compressors, or for standby diesel generation and diesel fire pumps.

The benefits may be listed as follows:

1. Common spare parts for different equipments mean that fewer need to be stocked and the weight of spares stored is reduced.
2. The reduced number of different types of equipment that operators have to deal with should reduce familiarization time and hence improve safety.
3. The total weight of equipment operation and maintenance manuals and design drawings on the average offshore installation is in excess of 10 tonnes, even when the majority is on microfiche; therefore some weight and space saving can also be accrued by reducing the total number of these.
4. There is a cost saving associated with items 1 to 3.

The offshore spares inventory must be optimized for items most critical to:

(a) safety;
(b) comfort of personnel (which may also affect safety);
(c) oil and gas production.

Small light items may be flown out quickly when required, but it should be remembered that if an unscheduled helicopter callout is required, it will cost £2000 or more (early 1990s).

14.2   Accessibility and communications

Those carrying out survey or commissioning work offshore should be warned that return travel from most if not all offshore installations requires customs clearance. This means that any test and measuring equipment needs proof-of-ownership documentation or a manifest for customs clearance.

When estimating travel time for such an offshore visit, some time should be allowed for weather delays. This is often worse in the summer months when fog is more likely.

Telephone communications are by line of sight and/or tropospheric scatter links, and access is usually available when major platform construction work is not in progress. On some of the nearer platforms, cellnet telephones have been put to good use to supplement normal communications.

14.3   Maintenance intervals and equipment specification

When offshore electrical equipment is being specified and procured, and in particular generator prime movers, manufacturers' recommended maintenance intervals should be carefully studied. Excessively short maintenance intervals will very quickly wipe out any savings in first cost, owing to the high costs of offshore servicing by manufacturers' or agents' servicing departments. Short intervals may also indicate poor reliability or unsuitability for the particular package, orientation or offshore environment in general.

14.4   Scaffolding

It is a common misconception that equipment or cabling may be accessed by scaffolding, almost invariably without major impact on cost or completion date. In some areas, such as a platform cellar deck or flare stack, the cost of erecting, maintaining and dismantling scaffolding may be the most significant cost in a small project.

Locating equipment and cables in inaccessible places, necessitating heavy use of scaffolding, has the following penalties:

1.   Heavy scaffolding requirements will discourage frequent maintenance, particularly if it is going to obstruct an accessway where traffic is heavy or an area which is already congested.
2.   The handling of scaffolding poles is well known in both onshore and offshore petrochemical installations for causing accidents or damage to process equipment.
3.   Scaffolding will also need to be erected each time it is necessary to inspect the equipment.
4.   Scaffolding offshore can be very costly. Costs in excess of £300 000 for a single access structure are not unknown.

Because of the need to access high-routed cable installations along their whole length in order to complete the installation, they are particularly expensive to install and should be avoided where possible. If an elevated or inaccessible location is unavoidable, consideration should be given to installing some form of permanent access structure, bearing in mind, of course, that this will itself need routine inspection, maintenance, painting etc.

14.5   Transport, accommodation and POB problems

When an offshore construction project is being planned, the limitations of helicopter transport and accommodation for personnel, crane lifting capacity, and storage/laydown space for equipment, play a large part in deciding the sequence of events.

Cranes are limited in lift capability by two main variables, namely radius of operation and sea state. If the crane is working near its maximum lift capacity, it is unlikely that the wave height in winter will allow lifting to be accomplished safely. Wind speed is also likely to affect the proceedings, but this will also depend on the wind direction and the location of the lifting operation with respect to this.

Ever since the sinking of the *Titanic* in 1912, it has been a statutory requirement that there are sufficient lifeboats for the number of people on board (POB). This applies just as much to fixed offshore installations as it does to the *QE2*, and must be borne in mind when planning a large offshore construction project. If necessary, owing to a shortfall in lifeboat capacity, a floatel accommodation vessel will need to be provided during project peak manning levels in order to comply with POB restrictions. This problem is often exacerbated by the need to send lifeboats ashore for maintenance during the summer months, when peak manning for platform maintenance is likely to occur. It can be overcome, however, by keeping a spare lifeboat of the same type onshore, which can be shipped out in advance to replace the one being serviced.

In general, the more detailed the documentation of the installation procedures during the design phase, the less likely it is that logistical problems similar to those outlined above will occur during construction.

Appendix A

Guide to offshore installations

A.1 Types of installation

A.1.1 Semisubmersibles

These are by far the most common type of installation, and are used as mobile drilling platforms, accommodation and construction centres and fire tenders. They are sometimes modified for use as oil production platforms. With a displacement often exceeding 20 000 tonnes, they float on buoyancy chambers placed well below the zone of wave action; and with a draft of up to 30 metres, they dampen the movement caused by the waves. The platforms are normally held in position by eight or more 15 tonne anchors and, like the fixed platforms, are designed to withstand extreme weather conditions.

A.1.2 Jackup or self-elevating platforms

A jackup rig is a barge-like vessel with steel legs each having racking gear which allows it to be racked down to the sea bed until the entire vessel is lifted out of the water. These vessels are obviously limited, by the length of the legs, to water depths of around 200 metres. They are inherently a temporary facility, more suited to drilling and maintenance activities than production.

A.1.3 Fixed platforms

A fixed platform may be described as consisting of two main components, the substructure and the superstructure.

A.1.3.1 The substructure
This is either a steel tubular jacket or a prestressed concrete structure.

Steel jacket structures normally consist of tubular legs held together by welded tubular bracing, the whole being securely piled to the sea bed through tubes attached to the bottom of the legs. Also in the structure are

258

various vertical steel tubes required for obtaining sea water for platform utilities (stilling tubes), for the protection of subsea electrical cabling (J-tubes), and as guiding for well risers. The jacket, when first installed without superstructure, may be more than 150 metres high.

Two proprietary designs of concrete gravity substructure are most common in the North Sea. The first is the C. G. Doris type, consisting of a large single watertight caisson surrounded by a circular structure containing a honeycomb of holes to act as a breakwater. A central column sprouts from the caisson, from which the superstructure is supported. The second is the Condeep type, consisting of circular towers or legs sprouting from a cellular caisson. In both cases, the weight of the structure ensures stability on the sea bed, avoiding the need for piles or anchors, and parts of the structure may be used for the storage of oil or housing production equipment.

A.1.3.2 The superstructure
The superstructure usually consists of a deck truss assembly which supports all the prefabricated production process facilities, drilling, utilities and accommodation modules, communication facilities, cranes etc. that make up the topsides of the platform.

With concrete platforms, it is normal to float out the substructure with as much of the superstructure in place as possible. This is to avoid the higher costs associated with carrying out installation work offshore.

Again, to minimize offshore installation work, which costs in the region of four times that of similar work onshore, all superstructure is modular. The modules are fitted out as completely as possible, taking into account the loading limits of available shipping and craneage capacity. Should it not be possible to complete the superstructure onshore then, when the modules have been landed, skidded into position and secured to the supporting deck assembly, the services and process connections such as electrical power, cooling water, HVAC ducting, process oil, condensate and gas may be interconnected between modules. This is known as the hookup phase.

A.1.4 Tension leg platforms

Where oil and gas fields are found in deeper water, the costs and technical problems involved in producing ever taller structures have forced designers to consider more novel methods.

The tension leg system is currently in use on one installation in the North Sea. It consists of a hull-like structure having positive buoyancy which is held to a deeper draught by means of tension legs at each corner. The legs consist of flexible metal tubes anchored in foundations in the sea bed and tensioned to hundreds of tonnes in the platform. A computer controlled ballasting system prevents load movements on the platform deck causing large differences in tension, and hence avoids dangerous stressing of the hull structure or the legs themselves.

A.2   Drilling essentials

A.2.1 Draw works

A rotating column of hollow steel pipe made up of sections screwed together, called a drill string, is fitted at the bottom with a cutting bit. It is hung from a swivel connection inside the derrick, a strong steel pylon usually 45 to 55 metres high. A wire cable runs through heavy pulleys to the draw works, which raise and lower the drill string. The draw works is driven by a DC motor variable speed drive controlled from the driller's console.

A.2.2 Rotary table

On the floor of the derrick is a rotary table with an opening in the middle through which slides the top section of the drill string, a square or hexagonal section pipe called the kelly. The rotary table is also driven by a DC motor variable speed drive controlled from the driller's console. The rotary table turns the kelly and hence the whole drill string. As the drill bit cuts deeper, extra sections are added to the drill string. The whole string will require to be lifted out to replace worn, damaged or inappropriate drill bits.

A.2.3 Mud pumps

During drilling operations, a special fluid is continuously pumped through the top of the kelly and down the hollow drill string to cool and clean the revolving bit below. The fluid also acts as a conveyor of drill cuttings to the surface and a means of sealing and supporting the wall of the hole.

The fluid is basically composed of a special clay suspended in water or oil, and is known as mud. Its composition must be varied according to conditions, and it may contain various additives for better lubrication of the drill bit in abrasive rock formations or better sealing in highly permeable rock formations.

The moving column of mud in the hole is vital to the safety of the drilling operation, since its weight helps to prevent a blowout if oil or gas at very high pressures is encountered. The petroleum engineer may increase the density of the mud at critical stages of drilling where high-pressure gas or oil pockets are likely. This increases the weight of the column of mud in the hole and reduces the blowout risk.

A.2.4 Shale shaker

The mud is circulated by mud pumps using yet another DC variable speed drive controlled from the driller's console. The drill cuttings conveyed to the surface are then sieved out by a vibrating screen called a shale shaker before returning to a pit for recirculation.

A.2.5 Blowout preventer

If a pocket of high-pressure oil or gas is unexpectedly pierced by the drill, the hydrostatic pressure and viscosity of mud in the drill hole will not be sufficient to prevent the oil and gas blowing the mud back and erupting from the well, often in a fierce jet. If this very hazardous condition, known as a blowout, is not quickly controlled, it may lead to serious fires and explosions, and possibly the loss of the entire installation.

To guard against blowouts, up to six heavy duty valves, collectively known as the blowout preventer, are fixed beneath the raised floor of the derrick or on the sea bed. If there is a risk of a blowout occurring, the well can be closed by these valves, even whilst drilling is in progress.

A.2.6 Wireline logging

One of the temporary facilities whose electrical supply is often problematical for electrical engineers is the wireline cabin. This is because it is almost invariably in the hazardous area created by the wellheads. It is necessary since a continuous log often needs to be made of the electrical and other properties of the rock formations drilled through. Different rock types, fluid content and physical properties can be distinguished, and with information obtained by analysis of drill cuttings and specially drilled core samples, a map of the oil or gas field geological structure may be built up.

A.2.7 Well completion

Once oil or gas is found in sufficient quantities to be economically viable, production wells may be drilled. In this case the hole is progressively lined with steel casing of different diameters. The production casing is the longest and has the smallest diameter, typically 245 mm, followed by intermediate casings and ending up with 760 mm surface casing.

The casings are inserted in a preplanned sequence as each section of the well is drilled, and are then cemented in place. If the producing layer is a firm one, such as limestone, the final part of the hole need not be lined and the oil can be produced directly into the bore well.

A.2.8 Perforation

If the producing layer is friable, a final string of casing sections is run right down to the bottom and cemented in place, and holes are shot through it with a perforating gun to let in the oil. This last operation usually requires radio silence to be maintained for several hours on the platform, as operation of transmitters can inadvertently trigger the perforating gun before it is in place.

If the producing formation is loose sand, a slotted pipe called a liner, with a gravel pack or wire filter round it, can be hung from the casing in the producing rock strata.

A.2.9 Christmas trees

At this stage, the well is still full of heavy drilling mud. Because the casing and liner must remain in a well for a long time and their repair or replacement would be expensive, another string of small-bore pipe, the tubing, hung from the wellhead at the top of the casing, is run down to the foot of the well, where it is sealed against the production casing with a packer. This is the pipe string through which the oil and gas flows to the surface.

The drilling mud is flushed out of the tubing with water, allowing the oil or gas to emerge. Owing to the high well pressures, often hundreds of bars, the oil flow needs careful regulation at the surface. A cluster of valves and fittings which seal off the well and control the flow, known as the Christmas tree, is a familiar sight in television documentaries about oil.

A.2.10 Subsea completions

Where other oil or gas fields are discovered adjacent to existing offshore production facilities, it may not be necessary to install a new platform or floating facility.

Instead, a sea bed wellhead facility is installed which, after production drilling, allows oil and gas to be piped to the adjacent established production facility for processing and export. The adjacent facility also acts as a control point for operation of the subsea wellhead valves and the point of supply for electrical equipment located on the subsea wellhead manifold.

A.2.11 Deviated drilling

With present day technology it is possible to guide the direction of drilling so that the well bore is deliberately deflected from the vertical in order to achieve good spacing at the reservoir to allow the gas and oil to be removed in a more uniform manner. The method may also be used where satellite wells are completed underwater. In some cases where the field is shallow, such as the Morecambe Bay gas field, wells may be drilled at an angle to the vertical using a special drilling rig with a derrick which may be tilted to the correct angle by hydraulic jacks before drilling commences.

A.3   Production process essentials

The processes outlined in the following sections are typical of a large North Sea fixed production platform.

A.3.1 Oil, water and gas separation

Petroleum mixtures are usually complex and not easy to separate efficiently. Vapour recovery and evaporation control must be engineered in a hostile environment with minimum risk of fire or explosion.

Separators on offshore installations consist of large steel pressure vessels, either horizontal or vertical, which, by a combination of many changes of flow direction and the action of gravity, cause the heavier liquids to fall to the bottom. After the first stage of separation, the gas will still be 'wet' and will require at least one further stage of separation. Liquids which would revert to a gas at atmospheric pressure are referred to as condensate, and they are normally re-entrained with the crude oil before it is exported from the platform by subsea pipeline or tanker. On some installations, the gas that can be liquefied is also injected into the oil pipeline before exporting ashore.

A large proportion of the flow from any North Sea well is water, and this must be removed from the oil and gas because it has no economic value and would also cause corrosion in the export subsea pipeline. As the petroleum liquids are less dense than water, they form a middle layer in the separators, so that the produced water can be drained off at the bottom. Level controllers ensure that the three different outputs of each separator are maintained. After further cleaning the water is normally dumped into the sea.

A.3.2 Hydrates

Mixtures of water vapour and natural gas at high pressure tend to form solid ice-like crystals on cooling, called hydrates. Hydrates can restrict or even completely prevent the flow of gas.

Some liquids have the property of absorbing water vapour from gases and are used to prevent or restrict the formation of hydrates. Ethylene glycol or similar agents are used for this purpose offshore. Gas is passed through a pressure vessel into which a fine spray of the agent is introduced.

A.3.3 Water injection

Sea water is drawn up by lift pumps and, sometimes with treated produced water, is injected into the oil bearing formations via a separate injection well. The purpose of this is to artificially maintain well pressure by replacing the oil, gas and water removed via the producing well. The injection pumps required are usually very large, of the order of several megawatts. The reinjection of produced water has the added advantage of disposing of a potential environmental pollutant, although with high oil prices it is economically attractive to centrifuge out quite low levels of entrained oil. Produced water re-injection is technically difficult to achieve, since very low levels of oil contamination will tend to increase injection back-pressure and severely restrict the available flow rate.

A.3.4 Gas reinjection

In the early days of oil production, it was accepted that the gas removed from the separators could be flared, should it be found uneconomical to pipe the gas ashore. Nowadays, with established North Sea gas pipe networks such as Far North Liquids and Associated Gas System Pipline (FLAGS), it is normally possible for the producing company to tie into an established gas line and supply the gas via its onshore terminal to a national gas utility. In any case, the flaring of large quantities of gas, except in abnormal circumstances, is strictly controlled by national regulations.

Where the gas export facilities of a platform are non-existent or are restricted to the extent that oil production would be affected, it is necessary to reinject the gas as an acceptable alternative to flaring. This also has the advantage of helping to maintain well pressure. Hopefully, the gas may be largely recoverable at a later date.

A large compressor train of two or three stages is normally required in order to obtain the gas flow rates and pressures necessary for reinjection. A typical scheme would be two stages utilizing axial compressors driven by large induction motors, followed by a third stage of a reciprocating compressor driven by a synchronous motor in order to ease voltage and frequency control problems for the platform generation. The problems of large process loads such as these, when powered from a small number of generators isolated from a national grid network, are discussed in Chapter 2.

A.3.5 Oil storage and export

Oil that has been through the platform processes may be exported immediately via large main oil line (MOL) pumps or, where no export pipeline exists, may be stored in the subsea cells of the platform structure prior to offloading by tanker. As the oil from the particular field may be above the average value of the oil from a shared export pipe network, it may be economically advantageous to export by tanker as well as by pipeline.

Where tanker loading is carried out, a separate floating tanker loading facility is necessary, as it would be highly dangerous for the tanker to approach within normal flexible pipe loading distance of the production platform. These single buoy moorings may have their own storage and pumping facilities and are connected to the associated platform via subsea pipelines.

A.3.6 Artificial lift facilities

The output from a typical oil field will rise to a maximum after all the production wells have been completed. There will then be a plateau in the output figures for, hopefully, a number of years before outputs start to decline. The timescales involved vary greatly with different fields and even with different wells in the same field, particularly where a lot of deviation drilling has been necessary. In order to maintain the oil output, submersible downhole pumps may be inserted down wells, or gas may be injected and released low down in the well in order to reduce the density of the oil. Both methods tend to increase oil flow rates and are known collectively as artificial lift facilities.

A.4   Oil company operations

A.4.1 Financing

As the cost of recovering oil from a field in the North Sea, from exploration to production, can easily exceed £1000 million, it is usually necessary for several oil companies and financial institutions to collaborate in the project. Typical costs and manpower requirements are shown in Table A.1.

It should be noted that the decline in output may be arrested or at least slowed by the use of artificial lift facilities. As the average cost of drilling a single exploration well is £3 million, a large part of £100 million may be spent on exploring a potential field before a decision is made to go ahead with the production facility.

Table A.1 Resources required for recovery of a typical oil field

	Exploration	*Construction*	*Production*
Programme	Survey of 2000 to 4000 km^2 Exploration and appraisal drilling of 5 to 30 wells	Planning/design Construction of platform structure, production and accommodation facilities Drilling of production wells Construction of transport facilities	Rising output 3–5 years Plateau 5 years Decline 8–10 years
Time	2–6 years	5–6 years	16–20 years
Direct labour	200–400 men	1000–2000 men	300–400 men
Capital outlay	£15–90 million	£750 million	£250 million depending on type of secondary recovery scheme (e.g. artificial lift) or other operations necessary
Operating costs			£250–500 million

A.4.2 Division of labour

When working on oil company projects, it is particularly necessary to be aware of the company management structure with respect to project funding, and therefore it is mentioned briefly here.

Most oil companies have four main management streams: operations, maintenance, projects and technical facilities. Those who take responsibility for the production plant, the asset holders, are the operators. The maintenance department may be refused access to an item of plant due for maintenance if it is still required by the operations department to maintain full oil production. The operations department may also refuse to accept new plant from a project construction team if they feel it is not performing adequately. The technical facilities department act only in an advisory capacity on matters of design and safety, although there is usually a separate safety department who carry out hazard and operability (hazop) studies and liaise with national inspectorates and shipping bureaux.

It is increasingly a practice for the larger oil companies to contract out the majority of their design and maintenance functions to a large engineering contractor.

Appendix B

Typical commissioning test sheets

PLATFORM:	FIELD:	PLANT NO:	SHEET OF
TAG NO:	SERIAL NO:	LOCATION:	SYSTEM:
TYPE:	P.O. NO:	DRG NO:	MANUFR:
VOLTAGE:	FREQUENCY:	RATING	DATE:
TESTER:	TEST COMPLETED:	SIGNED:	APPROVED:

SWITCHBOARDS AND BUSBARS

BUSBAR INSULATION RESISTANCE TEST RESULTS

BETWEEN	R-Y	R-B	Y-B	R-N	Y-N	B-N	RYB-E	N-E
IR (MEGOHMS)								

BUSBAR PRESSURE TEST RESULTS

BETWEEN	RY-B-E	RB-Y-E	RYB-E	RYB-N	N-E	
VOLTS AC/DC						
LEAKAGE AMPS						
DURATION MINUTES						

BUSBAR RESISTANCE MEASURED WITH A DUCTOR
SEE ATTACHED DRAWING OF BUSBARS SHOWING BUSBAR JOINTS

PHASE	MILLIOHMS	NO OF JOINTS	
RED			
YELLOW			
BLUE			
NEUTRAL			
EARTH PATH RESISTANCE			

VOLTAGES

CLOSING SUPPLY VOLTAGE		VOLTS
TRIPPING SUPPLY VOLTAGE		VOLTS

TEST EQUIPMENT

DEVICE	SERIAL NO	DEVICE	SERIAL NO

BUSBAR PHASE ROTATION ....

IR TEST ON ANTI-CONDENSATION HEATERS ....   MEGOHMS

267

PLATFORM:	FIELD:	PLANT NO:	SHEET OF
TAG NO:	SERIAL NO:	LOCATION:	SYSTEM:
TYPE:	P.O. NO:	DRG NO:	MANUFR:
VOLTAGE:	FREQUENCY:	RATING	
TESTER:	TEST COMPLETED:	SIGNED:	APPROVED:

POWER TRANSFORMER

AREA CLASSIFICATION	LOCATION
PRIMARY VOLTS	MANUFACTURER
SECONDARY VOLTS	RATING
COOLING	IMPEDANCE
VECTOR GROUP	FRQUENCY

WINDING RESISTANCE TEST RESULTS

BETWEEN	HV(R-Y)	HV(Y-B)	HV(B-R)	LV(R-Y)	LV(Y-B)	LV(B-R)
RESISTANCE						

WINDING INSULATION TEST RESULTS

	BEFORE	PRESSURE	TEST	AFTER	PRESSURE	TEST
BETWEEN	HV-EARTH	LV-EARTH	HV-LV	HV-EARTH	LV-EARTH	HV-LV
TEST VOLTS						
MEGOHMS						

WINDING PRESSURE TEST RESULTS

PRIMARY TEST VOLTAGE(kV)	DURATION (MINS)	LEAKAGE(mA)
SECONDARY TEST VOLTAGE(kV)		

MANUFACTURERS TEST CERTIFICATE	NO	POLARITY AND PHASE ROTATION	
OIL TEST CERTIFICATE	NO	TEMPERATURE TRIP OPERATIONAL	SETTING
RATIO AND MAGNETISING TEST CERT.	NO	TEMPERATURE ALARM OPERATIONAL	SETTING
FEEDER CIRCUIT BREAKER TEST CERT.	NO	OVER PRESSURE DEVICE	SETTING
HV AND LV CABLE TEST CERT.	NO		

TAP CHANGE MAGNETISING CURRENT RESULTS

	HV	WINDINGS		LV	WINDINGS	
	CONSTANT AT VOLTS					
	MAGNETIZING	CURRENT		SECONDARY	VOLTS	
TAP POSITION	R	Y	B	R-Y	Y-B	R-B
1						
2						
3						
4						
5						

PLATFORM:	FIELD:	PLANT NO:	SHEET OF
TAG NO:	SERIAL NO:	LOCATION:	SYSTEM:
TYPE:	P.O. NO:	DRG NO:	MANUFR:
VOLTAGE:	FREQUENCY:	RATING	
TESTER:	TEST COMPLETED:	SIGNED:	APPROVED:

INSPECTION TEST RECORD | CIRCUIT BREAKER SECONDARY INJECTION

CUBICLE REFERENCE

OVERCURRENT/EF RELAY		CT RATIO	
RELAY MANUFACTURER		TYPE	
SERIAL NUMBER		COIL RATING	
CHARACTERISTIC		PROTECTION	

INJ		SETTING		RELAY		TIME		RESET	
COIL	AMPS	PSM	TM	AMPS	ERROR %	CURVE	ACTUAL	AMPS	TIME
R									
Y									
B									
N									

CIRCUIT RATING					
FULL LOAD CURRENT					
SETTING EF		PSM		TM	
SETTING OC		PSM		TM	

MOTOR PROTECTION RELAY

RELAY MANUFACTURER		TYPE	
SERIAL NUMBER		COIL RATING	
CHARACTERISTIC		PROTECTION	

TEST	AMPS	PLUG SET	CURVE	ACTUAL	RESET (SECS)
START CURVE					
RUNNING CURVE					
RUNNING CURVE					
EARTH FAULT					
INSTANTANEOUS TRIP					

FULL LOAD CURRENT		% LOAD TO TRIP	
PLUG SETTING		INST TRIP	

COMMENTS/EXCEPTIONS

PLATFORM:	FIELD:	PLANT NO:	SHEET OF
TAG NO:	SERIAL NO:	LOCATION:	SYSTEM:
TYPE:	P.O. NO:	DRG NO:	MANUFR:
VOLTAGE:	FREQUENCY:	RATING	
TESTER:	TEST COMPLETED:	SIGNED:	APPROVED:

INSPECTION TEST RECORD	ELECTRIC MOTOR

MOTOR NUMBER			
AREA CLASSIFICATION			
FRAME SIZE		CONNECTION STAR ☐ DELTA ☐	
LUBRICATION		ROT VIEWED FROM DE CW ☐ CCW ☐	
VOLTAGE	SPEED	R.P.M.	CURRENT F.L.C.	AMPS
RATING	KW	FREQUENCY	H.Z.	O/L SETTING	AMPS
ED FROM			

APPROPRIATE FORM COMPLETED FOR EX MOTORS ....

.... SIGNED .... DATE ....

WINDING RESISTANCE ....

POLARISATION INDEX .... MEGOHMS AFTER 1 MIN .... MEGOHMS AFTER 10 MINS ....

BEARING INSULATION RESISTANCE ....

ALL ALARM CIRCUITS TESTED ....

REMOTE CONTROL CIRCUITS TESTED ....

ANTI-CONDENSATION HEATER CIRCUIT TESTED ....

CIRCUIT BREAKER/CONTACTOR TEST CERTIFICATES AVAILABLE ....

CABLES TEST CERTIFICATES AVAILABLE ....

TIME	INSULATION RESISTANCE	3-PHASE CURRENT IN AMPS			BEARING TEMPERATURE		WINDING TEMPERATURE	MOTOR COOLING	MOTOR SPEED	VIBRATION
	MEGOHMS	B	W	R	FRONT	REAR			R.P.M.	
UNCOUPLED										
ON LOAD										

COMMENTS/EXCEPTIONS

CONTACTOR STARTER	INSPECTION TEST RECORD

PLATFORM:	FIELD:	PLANT NO:	SHEET OF
TAG NO:	SERIAL NO:	LOCATION:	SYSTEM:
TYPE:	P.O. NO:	DRG NO:	
VOLTAGE:	FREQUENCY:	RATING	
TESTER:	TEST COMPLETED:	SIGNED:	APPROVED:

MANUFACTURER		COMPARTMENT No	
MCC No.		DRIVE ALLOCATION	
OVERCURRENT TRIP SETTING	AMPS	TIMER T . . . . . SETTING	SECS
DRIVE FULL LOAD CURRENT	AMPS	CONTROL FUSE SIZE	AMPS
MOTOR FUSE SIZE	AMPS	CONTROL VOLTAGE	VOLTS
SYSTEM VOLTAGE	VOLTS		

SIGNED . . . . DATE . . . .

INSULATION RESISTANCE . . . . MEGOHMS   TEST VOLTAGE . . . . VOLTS

CONTROL CIRCUITS CORRECT TO SCHEMATIC DIAGRAM No . . . .   REV . . . .

REMOTE CONTROL AND ANNUCIATOR TESTED AND CORRECT . . . .

ANTI-CONDENSATION HEATER CIRCUIT TESTED . . . .

THERMAL OVERLOAD TEST RESULTS				
CURRENT		TRIP TIME (SECS)		REMARKS
% FLC	AMPS	R	Y	B	
115					
					NO TRIPS
					MIN. TRIP AMPS
200					
600					

CT RATIO PROTECTION . . . .

CT RATIO METERING . . . .

SINGLE PHASING TEST RESULTS (80% FLC)			
CIRCUIT	R-Y	R-B	Y-B
AMPS			
TRIP TIME (SECS)			

EARTH FAULT TEST RESULTS			
PHASE	R	Y	B
MIN. AMPS TO OPERATE			

AMMETER TEST RESULTS	
LOCATION TEST AMPS	
READING	
READING	
READING	

COMMENTS/EXCEPTIONS

Appendix C

Simple three-phase fault calculation program

C.1 Program listing

```
10      REM This program is written in GWBASIC and should run on any PC
20      REM or compatible. The program is complicated by the lack of imaginary
30      REM number handling in GWBASIC and would be simpler written in FORTRAN.
40      DEFINT I-N
50      DIM C1 (20,20), C2 (20,20), Y4 (20,20), Y5 (20,20)
60      DIM Y1 (20,20), Y2 (20,20), YY1 (20,20), YY2 (20,20)
70      DIM V1 (20), V2 (20), F5 (20), F6 (20), C3 (20), C4 (20), C5 (20)
80      DIM FL (20), TA$ (20), C7 (20)
90      INPUT"CONTRACT NO.                                              : ", WA$
100     INPUT"CLIENT                                                    : ", CA$
110     INPUT"LOCATION                                                  : ", TA$
120     INPUT"REFERENCE DRW. NO.                                        : ", RA$
130     INPUT"CLIENT REFERENCE                                          : ", CB$
140     INPUT"SYSTEM MVA BASE                                           : ", BASE
150     INPUT"TOTAL NUMBER OF BUSBARS                                   : ", NB
160     INPUT"TOTAL NUMBER OF INTERCONNECTORS                           : ", NL
170     FOR I=1 TO NB
180     FOR J=1 TO NB
190     Y1 (I, J) =0!
200     Y2 (I, J) =0!
210     YY1 (I, J) =0!
220     YY2 (I, J) =0!
230     Y4 (I, J) =0!
240     Y5 (I, J) =0!
250     NEXT J
260     NEXT I
270     FOR I=1 TO NB
280     PRINT USING"BUS NO. ##"; I;
290     INPUT" VOLTAGE BASE IN VOLTS                         : ", VBASE (I)
300     NEXT I
310     FOR I=1 TO NL
320     PRINT USING"LINE NO. ## DATA : "; I
330     INPUT"                            START BUS             : ", L
340     INPUT"                            END BUS               : ", M
350     INPUT"                            Resistance in p.u. : ", R
360     INPUT"                            Reactance in p.u.  : ", X
370     XA=-R/ (R^2+X^2)
380     XB=X/ (R^2+X^2)
390     IF L=M THEN 450
400     Y1 (L, M) =XA
410     Y2 (L, M) =XB
420     Y1 (M, L) =XA
430     Y2 (M, L) =XB
440     GOTO 490
450     YD1=XA
460     YD2=XB
```

272

```
470      Y1 (L, M) =Y1 (L, M) +YD1
480      Y2 (L, M) =Y2 (L, M) +YD2
490      NEXT I
500      FOR I=1 TO NB
510      PRINT USING"Busbar No. ## p.u. Voltage Data :"; I
520      INPUT"                              Real part Voltage (p.u.)   : ", V1 (I)
530      INPUT"                              Imag'y part Voltage (p.u.)  : ", V2 (I)
540      FOR J=1 TO NB
550      Y4 (I, I) =Y4 (I, I) −Y1 (I, J)
560      Y5 (I, I) =Y5 (I, I) −Y2 (I, J)
570      IF I=J THEN 600
580      Y4 (I, J) =Y1 (I, J)
590      Y5 (I, J) =Y2 (I, J)
600      NEXT J
610      NEXT I
620      LPRINT: LPRINT: LPRINT
630      LPRINT TAB (10) "****************************************************
********"
640      LPRINT TAB (10) "*" TAB (70) "*"
650      LPRINT TAB (10) "*" TAB (26) "-- FAULT LEVEL CALCULATION --" TAB (70) "*"
660      LPRINT TAB (10) "*" TAB (70) "*"
670      LPRINT TAB (10) "*          CONTRACT NO.           "; TAB (55) ; WA$; TAB (70) ; "*"
680      LPRINT TAB (10) "*          CLIENT                 "; TAB (55) ; CA$; TAB (70) ; "*"
690      LPRINT TAB (10) "*          LOCATION               "; TAB (55) ; TA$; TAB (70) ; "*"
700      LPRINT TAB (10) "*          REFERENCE DRAWING NO. "; TAB (55) ; RA$; TAB (70) ; "*"
710      LPRINT TAB (10) "*          CLIENT REFERENCE       "; TAB (55) ; CB$; TAB (70) ; "*"
720      LPRINT TAB (10) "*" TAB (70) "*"
730      LPRINT TAB (10) "****************************************************
********"
740      LPRINT: LPRINT: LPRINT: LPRINT
750      GOSUB 1150
760      FOR I=1 TO NB
770      FOR J=1 TO NB
780      YY1 (I, J) =C1 (I, J)
790      YY2 (I, J) =C2 (I, J)
800      NEXT J
810      NEXT I
820      FOR I=1 TO NB
830      C1K=YY1 (I, I) / (YY1 (I, I) ^2+YY2 (I, I) ^2)
840      C2K=−YY2 (I, I) / (YY1 (I, I) ^2+YY2 (I, I) ^2)
850      C1J=−V1 (I) *C1K+V2 (I) *C2K
860      C2J=−V2 (I) *C1K−V1 (I) *C2K
870      C3 (I) =C1J
880      C4 (I) =C2J
890      C7 (I) =SQR (C3 (I) ^2+C4 (I) ^2) * (BASE*1000!) / (SQR (3) *VBASE (I))
900      C5 (I) =SQR (C3 (I) ^2+C4 (I) ^2)
910      FL (I) =ABS (C5 (I) *BASE)
920      LPRINT
930      LPRINT USING" -- FOR FAULT OCCURRENCE AT BUS NO. ## --"; I
940      LPRINT
950      LPRINT USING" SYMMETRICAL SHORT-CIRCUIT CURRENT = ###.#### kA"; C7 (I)
960      LPRINT
970      LPRINT USING" FAULT LEVEL                          = ####.### MVA"; FL (I
980      FOR J=1 TO NB
990      IF I=J THEN 1110
1000     C1K=YY1 (I, I) / (YY1 (I, I) ^2+YY2 (I, I) ^2)
1010     C2K=−YY2 (I, I) / (YY1 (I, I) ^2+YY2 (I, I) ^2)
1020     C1J=YY1 (J, I) *C1K−YY2 (J, I) *C2K
1030     C2J=YY2 (J, I) *C1K+YY1 (J, I) *C2K
1040     C1K=C1J*V1 (I) +C1J*V2 (I)
1050     C2K=C2J*V1 (I) +C1J*V2 (I)
1060     F5 (J) =V1 (J) −C1K
1070     F6 (J) =V2 (J) −C2K
1080     LPRINT
1090     C6=SQR (F5 (J) ^2+F6 (J) ^2)
1100     LPRINT USING" FAULT VOLTAGE ON BUSBAR NO. ##     = ###.### p.u"; J; C6
1110     NEXT J
1120     LPRINT: LPRINT: LPRINT
```

```
1130    NEXT I
1140    END
1150    N=NB
1160    FOR I=1 TO N
1170    FOR J=1 TO N
1180    C1(I,J)=Y4(I,J)
1190    C2(I,J)=Y5(I,J)
1200    NEXT J
1210    IJ(I)=0!
1220    NEXT I
1230    ITER=0!
1240    ITER=ITER+1
1250    IF ITER>N THEN 1660
1260    CNV=0
1270    FOR J=1 TO N
1280    IF IJ(J)>0 THEN 1340
1290    CA=SQR(C1(J,J)^2+C2(J,J)^2)
1300    TEST=ABS9CA)
1310    IF TEST<CNV THEN 1340
1320    CNV=TEST
1330    K=J
1340    NEXT J
1350    IJ(K)=1
1360    C1K=C1(K,K)/(C1(K,K)^2+C2(K,K)^2)
1370    C2K=-C2(K,K)/(C1(K,K)^2+C2(K,K)^2)
1380    C1(K,K)=C1K
1390    C2(K,K)=C2K
1400    FOR I=1 TO N
1410    FOR J=1 TO N
1420    IF (I-K)=0 THEN 1520
1430    IF (J-K)=0 THEN 1520
1440    C1K=C1(K,K)*C1(K,J)-C2(K,K)*C2(K,J)
1450    C2K=C2(K,K)*C1(K,J)+C1(K,K)*C2(K,J)
1460    C1J=C1(I,K)*C1K-C2(I,K)*C2K
1470    C2J=C2(I,K)*C1K+C1(I,K)*C2K
1480    C1K=C1(I,J)-C1J
1490    C2K=C2(I,J)-C2J
1500    C1(I,J)=C1K
1510    C2(I,J)=C2K
1520    NEXT J
1530    NEXT I
1540    FOR I=1 TO N
1550    IF (I-K)=0 THEN 1640
1560    C1K=C1(I,K)*C1(K,K)-C2(I,K)*C2(K,K)
1570    C2K=C2(I,K)*C1(K,K)+C1(I,K)*C2(K,K)
1580    C1(I,K)=C1K
1590    C2(I,K)=C2K
1600    C1K=-C1(K,I)*C1(K,K)*C2(K,I)*C2(K,K)
1610    C2K=-C2(K,I)*C1(K,K)-C1(K,I)*C2(K,K)
1620    C1(K,I)=C1K
1630    C2(K,I)=C2K
1640    NEXT I
1650    GOTO 1240
1660    FOR I=1 TO N
1670    FOR J=1 TO N
1680    NEXT J
1690    NEXT I
1700    RETURN
1710    END
```

Program courtesy of Dr Tsao Ta'pang of the University of Sun Yat-Sen ROC.

Note: This program is for illustration purposes only. It will not calculate earth faults as no algorithm has been included for handling the different transformer winding arrangements. Motor fault contributions are not included. Please refer to the commercial programs available, listed in Appendix D.

C.2   Example input

Input may be organized using Figure C.1.

CONTRACT NO.   : C 56−D23
CLIENT   : OILCO
LOCATION   : PLANKTON ALPHA
REFERENCE DRW. NO.   : Q/GG/456
CLIENT REFERENCE   : 22−897
SYSTEM MVA BASE   : 10
TOTAL NUMBER OF BUSBARS   : 4
TOTAL NUMBER OF INTERCONNECTORS   : 9
BUS NO. 1 VOLTAGE BASE IN VOLTS   : 6600
BUS NO. 2 VOLTAGE BASE IN VOLTS   : 440
BUS NO. 3 VOLTAGE BASE IN VOLTS   : 440
BUS NO. 4 VOLTAGE BASE IN VOLTS   : 440
LINE NO. 1 DATA:
                              START BUS   : 1
                              END BUS   : 1
                              Resistance in p.u.   : 0
                              Reactance in p.u.   : 0.02378

LINE NO. 2 DATA:
                              START BUS   : 1
                              END BUS   : 2
                              Resistance in p.u.   : 0
                              Reactance in p.u.   : 0.002759
LINE NO. 3 DATA:
                              START BUS   : 1
                              END BUS   : 3
                              Resistance in p.u.   : 0
                              Reactance in p.u.   : 0.552
LINE NO. 4 DATA:
                              START BUS   : 1
                              END BUS   : 4
                              Resistance in p.u.   : 0
                              Reactance in p.u.   : 0.552
LINE NO. 5 DATA:
                              START BUS   : 2
                              END BUS   : 2
                              Resistance in p.u.   : 0
                              Reactance in p.u.   : 2.05
LINE NO. 6 DATA:
                              START BUS   : 3
                              END BUS   : 3
                              Resistance in p.u.   : 0
                              Reactance in p.u.   : 2.469
LINE NO. 7 DATA:
                              START BUS   : 4
                              END BUS   : 4
                              Resistance in p.u.   : 0
                              Reactance in p.u.   : 2.05
LINE NO. 8 DATA:
                              START BUS   : 2
                              END BUS   : 3
                              Resistance in p.u.   : 0.1456
                              Reactance in p.u.   : 0.27
LINE NO. 9 DATA:
                              START BUS   : 3
                              END BUS   : 4
                              Resistance in p.u.   : 0.105
                              Reactance in p.u.   : 0.213
Busbar No. 1 p.u. Voltage Data:
                    Real part Voltage (p.u.)   : 1.02
                    Imag'y part Voltage (p.u.)   : 0
Busbar No. 2 p.u. Voltage Data:
                    Real part Voltage (p.u.)   : 1
                    Imag'y part Voltage (p.u.)   : 0
Busbar No. 3 p.u. Voltage Data:
                    Real part Voltage (p.u.)   : 1
                    Imag'y part Voltage (p.u.)   : 0
Busbar No. 4 p.u. Voltage Data:

276

1. Base MVA =	2. Base voltage (volts) =
3. Number of busbars =	4. No. of interconnection links =
5. Busbar voltages in volts	NOTE: This number should include the self-impedances entered

Busbar no.	Voltage (volts)	Busbar no.	Voltage (volts)
1		9	
2		10	
3		11	
4		12	
5		13	
6		14	
7		15	
8			

Sheet 1
Figure C.1 Data input sheet for fault calculation program

7. Driving voltages

Busbar no.	Real pu volts	1 mag pu volts
1		
2		
3		
4		
5		
6		
7		
8		

Busbar no.	Real pu volts	1 mag pu volts
9		
10		
11		
12		
13		
14		
15		

NOTE: If send and receive bus numbers are identical, the impedance shown is the equivalent self-impedance corresponding to the total fault contribution on the switchboard from external sources and/or rotating plant.

Send bus	Receive bus	Xpu	Rpu
1			
2			
3			
4			
5			
6			
7			
8			

Send bus	Receive bus	Xpu	Rpu
9			
10			
11			
12			
13			
14			
15			

Sheet 2

C.3 Example results

```
*********************************************************************
*                     — FAULT LEVEL CALCULATION —
*          CONTRACT NO.                          C 56-D23
*          CLIENT                                OILCO
*          LOCATION                              PLANKTON ALPHA
*          REFERENCE DRAWING NO.                 Q/GG/456
*          CLIENT REFERENCE                      22-897
*********************************************************************
```

-- FOR FAULT OCCURRENCE AT BUS NO. 1 --

SYMMETRICAL SHORT-CIRCUIT CURRENT = 73.0950 kA

FAULT LEVEL   = 835.587 MVA

FAULT VOLTAGE ON BUSBAR NO. 2   =   0.569 p.u.

FAULT VOLTAGE ON BUSBAR NO. 3   =   0.103 p.u.

FAULT VOLTAGE ON BUSBAR NO. 4   =   0.641 p.u.

-- FOR FAULT OCCURRENCE AT BUS NO. 2 --

SYMMETRICAL SHORT-CIRCUIT CURRENT = 311.4576 kA

FAULT LEVEL   = 237.363 MVA

FAULT VOLTAGE ON BUSBAR NO. 1   =   1.020 p.u.

FAULT VOLTAGE ON BUSBAR NO. 3   =   0.535 p.u.

FAULT VOLTAGE ON BUSBAR NO. 4   =   0.363 p.u.

-- FOR FAULT OCCURRENCE AT BUS NO. 3 --

SYMMETRICAL SHORT-CIRCUIT CURRENT = 114.0009 kA

FAULT LEVEL   = 86.880 MVA

FAULT VOLTAGE ON BUSBAR NO. 1   =   1.020 p.u.

FAULT VOLTAGE ON BUSBAR NO. 2   =   1.007 p.u.

FAULT VOLTAGE ON BUSBAR NO. 4   =   0.285 p.u.

-- FOR FAULT OCCURRENCE AT BUS NO. 4 --

SYMMETRICAL SHORT-CIRCUIT CURRENT = 55.4643 kA

FAULT LEVEL   = 42.269 MVA

FAULT VOLTAGE ON BUSBAR NO. 1   =   1.020 p.u.

FAULT VOLTAGE ON BUSBAR NO. 2   =   1.003 p.u.

FAULT VOLTAGE ON BUSBAR NO. 3   =   0.971 p.u.

Commercial programs for load flow, fault and transient stability

ERA Technology Ltd, Cleeve Road, Leatherhead, Surrey KT22 7SA, UK. Program: ERACS.

Service in Information and Analysis Limited (SIA), Ebury Gate, 23 Lower Belgrave Street, London SW1W 0NW, UK. Program: IPSA (developed by UMIST)

Cyme International Inc., 1485 Roberval, Suite 204, St Bruno, Quebec, Canada J3V 3PB. Program: CYMSTAB

Central Electricity Generating Board, Planning Department: System Technical Branch, Sudbury House, 15 Newgate Street, London EC1A 7AU, UK. Program: RAS MO5

ABB Network Control, Department NSN, 5300 Turgi, Switzerland. Program: MANISTA-386

Note that the programs supplied are modular, and usually require load flow and other data processing programs to be run prior to the transient stability program in order to establish initial conditions.

Appendix E Comparison of world hazardous area equipment

Marking systems, comparison of national and international standards and symbols for explosion protection. Table by H. Dreier (PTB) Jan. 1973. Revised for Middle East Electricity, August 1979. Reproduced with the permission of R. Stahl, Künzelsau.

Country, organisation, legislation	Specifications for construction and test	Specifications for electrical installations	Method of Wiring: cable (type 'e') conduit (type 'd')	General symbol for explosion-protection	I Flameproof/explosion-proof enclosure	Expl.-group	Symbols used for: Sandfilling	Increased safety	Pressurizing (purging)	Oil-immersion	Intrinsic safety	Special protection	Dust-explosion-protection	Temperature-classes	Classification of areas	Testing authorities and general remarks
IEC International Electrot. Com.	IEC-P.[b]. 79-1..11	Under review TC31/WG9	cab + con	Ex	d	I, IIA-C	q	e	p	o	ia, 1b	(S)	under review TC31H	T1-T6	0,1,2	—
CI C "EC-Directive" CENELEC EN.	European standards	In preparation acc. to IEC	cab	Ex EEx	d	I, IIA-C	q	e	p	o	ia, 1b	s		T1-T6	·	(Stand. Prüfst.-Konf.) EG
D W. Germany VDE DKE PAVO	Mining: VDE 0170 Others: VDE 0171	VDE 0118 VDE 0165	cab	(Sch) (Ex)	d	1-3n	(s)	e	f/p	o	I	s		G1-G5	0,1,2	BVS Dortmund-Derne PTB Braunschweig
F France U.T.E. Decret 61-295	NF C 12-300/320 C 20-061 C 23-210 (-S.A.)		con	MS AE	ADF	GI. GII-IV	RD	SA	SP		SI			- (200°C)	E. F. G	CERCHAR. Verneuil-en-Halette L.C.I.E. Fontenay-aux-Rose
I Italy C.E.I.	CEI 31-1 (X-1969) only for AD-PE	CEI 64-2 (ab 7.73)	con	AD Ex-	PE d	IIA-C Cl.1	S q	FE e	SI p	S o	I I	S	Cl.2	T1-T6	0,1,2	CESI. Milano. AD for Installations Ex for apparatus
B Belgium C.E.B.	d: NBN 286 e: NBN 717 I: NBN 683	Regl. Gén. p: NBN 716	cab + con	Ex	ADF	I, IIa-c 2c-f	(s)	e (SA)	(SI)		(I)			N. O. P. Q. G1-G5	0,1,2 and 3	INIEX. Paturages
NL Netherlands REGO	NEN 3150 NEN 3125/69	NEN 1010	cab + con	Ex	F	IIA-C	Q	E	P	O	I	H		T1-T6	4 Zo.	(formerly KEMA, Arnhem)
GB United Kingdom B.S.I. Fact. Act'61	d: BS 229 I: BS 1259 BS 4683	CP 1003/1 & 2 IP El. Saf. Code' 63	cab + con	(Ex)	FLP d	Gr.I. II-IV IIA-C		e	p		IS I			- (200°C) T1-T6	0,1,2	S.M.R.E. Buxton/Sheffield BASEEFA. Buxton (Certif. Standards)

DK	Denmark Heavy Current Regs.	Afsnitt 33	Afsnitt 7 (L...0)	cab	Ex Ex300	d	L. II-IVn		H/N	f/fu	o	N/I	s	<300°C ≥300°C		DEMKO. Herlev
D	E. Germany	TGL-19491 ...19503 (since 1.1.73)	TGL-200-0621	cab	(Sch) (Ex)	d / d	I...0		e		o		s	T1–T6	B-lab -IId	Institut f.Bergbausicherheit Leipzig, Bereich Freiberg
SU	USSR	GOST...	PLE VII-3	cab + con	PBa	PBa B-V	1 / 2-4	K	H / N	II / P	M / M	9	C / S	(A.B.G.D) T1–T5	2.1.0	SIPEccA. Donetsk: OWFI
CS	Czechoslovakia	CSN 341480 / CSN 341499	CSN 341440	cab	(Ex)	3	M / P.S.H	1	0	6	5.	9	8	A–F (T1–T6)		Prüfstelle Nr. 214, co VVUU. Ostrava-Radvanice
H	Hungary	MSZ 4814/	MSZ1600/8	cab + con	Sh- / Rh-	n	II-Ivx	h	f	t	c	sz	k	G1–G5		BKI. Budapest
PL	Poland	PN-72/E-08110	PN-63/E-05050 / PBUE	cab + con	B / Ex	M	I / IIA-C	Z	W	P	O	J	S	T1–T6	–	Vers.-"Grube "Barbara" (KDB). Mikolow
YU	Yugoslavia S-Commission. Zagreb. J.ZS.. Beograd	JUS Group N.S8	Install. regulations SL.L 18/67	cab	Ⓢ (S..)	t	L. / IIA-D	q	s	p	o	I	n	T1–T6	0.1.2	Electrotechn. Institute "Rade Koncar". Zagreb
A	Austria Elektr. Ges.	ÖVE 171	ÖVE 165	cab	Ex	d	1-3n		e	f	o	I	s	G1–G5		ETVA. Vienna / TÜV. Vienna
N	Norway NVE	similar to VDE 0171	similar to VDE 0165 art. 495	cab	Ex / Ⓝ	d	1-3n		e	f/p	o	I	s	G1–G5	(O) 1-IIa.b (2)	NEMKO. Oslo. Approval by Sprengstoffinspektionen. Tonsberg
S	Sweden	SEN 21 08.. (9.'69)		cab	x / SP	t	1-3		h	v	o	I	s	T1–T5		Statens Provningsanstalt (SP) Boras
CH	Switzerland	SEV 1015		cab + con	(Ex)	d	1-3		e	f	o	I	s	A–D		SEV. Zurich
J	Japan	JIS...	Rec. Pract. '65 (engl)	con / MIC		d	1-3n		e	f	o	I	s	G1–G5	0.1.2	Res. Inst. of Industrial Safety. Tokyo. (RIIS)
USA	U.S.A. Underwriters FMRC OSHA	UL Stand. for Safety No674-913	NEC 500 / ISA RP12.. / API RP 550		Cl.I (Gas)	ex- pl.- proof	D-A					IS		Cl.II +III Gr.E.F. G	1.2	UL Lab's. Northbrook / FMRC. Norwood/Mass. (NIOSH for Bureau of Mines)
CDN	Canada CSA	CSA Std. C22.2 No.30 (Enclosures) C22.2 No.145 (Motors)	CSA Std. C22.1 (1rd.) (for mines)	con	Ⓢ Ind.	Cl.I	D-A							Cl.II Gr.E.F. G / T1–T6	1.2	Can. Explosive Atm. Lab. Fuels Res. Centre. Ottawa. CSA Testing Laboratories. Toronto (Except for fdp)

Examples of the marking of explosion-protected electrical equipment in the type of protection 'Flameproof Enclosure' according to their groups (explosion-classes) related to German VDE classification.

Explosion class (VDE 0170/0171)	Typical gases	1 z.B. Methane CH_4 Firedamp CH_4 (Mining)	1 z.B. Petrol $C_n H_{n+2}$	2 z.B. Ethylene $CH_2=CH_2$	3a 3b 3c Hydrogen H_2, Carbon disulphide CS_2, Acetylene $CH\equiv CH$	3n All Gases
IEC	Int. Electr. com.	Ex d I	Ex d IIA	Ex d IIB	Ex d IIC	
D	W. Germany	(Sch)d	(Ex)d1	(Ex)d2	(Ex)d3a, 3b, 3c	(Ex)d3n
F	France	ADF GI	ADF GII	ADF GIII	ADF GIV	ADF GIV
B	Belgium	Ex I	Ex IIa	Ex IIb	Ex IIc	Ex IIc
NL	Netherlands, older	I IM	I II	I III	I IV	I IV
	Netherlands NEN 3125		Ex F IIA	Ex F IIB	Ex F IIC	Ex F IIC
GB	United Kingdom	Ex FLP Gr. I	Ex FLP Gr. II	Ex FLP Gr. III	Ex FLP Gr. IV	Ex FLP Gr. IV
D	E. Germany	(Sch)dI	(Ex)dII	(Ex)dIII	(Ex)dIVa, IVb, IVc	(Ex)dIVn
SU	USSR	B 1.	B 2	B 3	B 4	B 4
CS	Czechoslovakia	Ex 3 M	Ex 3 P	Ex 3 S	Ex 3 H_2, CS_2, C_2H_2	Ex 3 H
H	Hungary	Sb-n I	Rb-n II	Rb-n III	Rb-n IVa, IVb, IVc	Rb-n IVx
PL	Poland	BM	Ex-M IIA-	Ex-M III	Ex-M IVa, IVb, IVc	Ex-M IVn
YU	Yugoslavia	S t I	S t IIA	S t IIB	S t II C......	S t IIC
A	Austria	(Sch)d	(Ex)d1	(Ex)d2	(Ex)d3a, 3b, 3c	(Ex)d3n
S	Sweden	—	X t 1	X t 2	X t 3	X t 3
CH	Switzerland	—	(Ex)d1	(Ex)d2	(Ex)d3	(Ex)d3
USA	United States	Cl. I Gr. D	Cl. 1 Gr.C	Cl. 1 Gr.C	Cl. 1Gr. B. B. A	Cl. 1 Gr.A
CLC	European Standards (EN)	EEx d I	EEx d IIA	EEx d IIB	EEx d IIC (IIB + H_2 or IIB + CS_2)	EEx d IIC

Bibliography

Text sources

General

Crook, Leo *Oil Terms: Dictionary of Terms Used in Oil Exploration and Development*

Dunphy, Elaine *Oil: A Bibliography* (5th edn, microfiche only), Aberdeen and North of Scotland Library and Information Cooperative Service (ANSLICS),

Gunter, G. Siep (ed.) *Siemens Electrical Installation Handbook* (3 vols), Siemens AG and Wiley, 1987

Holtzer, J. M. (ed.) *Elsevier's Oilfield Dictionary*, Elsevier

Chapter 1

Crawford, J. *Offshore Installation Practice*, Butterworths

Chapter 2

Cullen, the Hon. Lord *The Public Enquiry into the Piper Alpha Disaster* (2 vols), HMSO

Chapter 3

Det Norske Veritas *Gas Turbines: Ventilation and Area Classification*, Technical Note for Fixed Offshore Installations

Engineering Equipment & Material Users Association/Institute of Petroleum, *Recommendations for the Protection of Diesel Engines Operating in Hazardous Areas* (MEC 1)

IEC 363 (*see list of standards below*)

Chapter 4

Chalmers, B. J. *Electric Motor Handbook*, Butterworth
National Fire Protection Association (US) *Diesel-Electric Fire Pumps*, specification NFPA 20

Chapter 5

American Petroleum Institute *Manual on Installation of Refinery Instruments and Controls. Part 1: Process Instrumentation and Control*, Recommended Practice RP 550
BS 6133: 1985 'Safe operation of lead-acid stationary cells and batteries'

Chapter 6

Cooper, C. B., McLean, D. M. and Williams, K. G. 'Application of test results to the calculation of short circuit fault levels in large industrial systems with concentrated induction motor loads', *Proc, IEE*, 1969, **116** (11), pp.
Lythall, R. J. *The J & P Switchgear Book* (ed. C. A. Worth), Butterworths

Chapter 7

BS 5750: Parts 1 and 2: 1987 (*see list of standards below*)

Chapter 8

Institute of Petroleum (UK) Model Code of Safe Practice
   Part 1 *Electrical Safety Code*, 1965
Part 15 *Area Classification Code for Petroleum Installations*, 1990
BS 5420: 1977 (1988) (*see list of standards below*)
BS 4999: Part 105: 1988 (*see list of standards below*)

Chapter 9

BS 142: 1982 (*see list of standards below*)
BS 3938: 1972 (1982) (*see list of standards below*)
BS 3950: 1965 (*see list of standards below*)
Bergen, Arthur R. *Power Systems Analysis*, Prentice-Hall, 1986
Davies, T. *Protection of Industrial Power Systems*, Pergamon Press, 1984
Electricity Council Power System Protection (3 Vols.) Peter Peregrinns Ltd.
Fisher, Lawrence E. 'Resistance of Law Voltage Arcs,' *IEEE Trans.* IGA-6, Nov/Dec, 1970
GEC Measurements *Protective Relay Application Guide* (3rd edn), 1987
General Electric Company (US) *Protective Relays Reference Data*, publication GEZ-7278

Westinghouse Electric Corporation *Electrical Transmission and Distribution Reference Book*

Chapter 10

BS 4211: 1987 (*see list of standards below*)
BS 5266: 1981 *et seq.* (*see list of standards below*)
CIBSE *Lighting Design Guide*
CIBSE *Application Guide: Lighting in Hostile and Hazardous Environments*, 1983
CIBSE *Industrial Area Floodlighting*, Technical Report TR13
CIBSE *The Calculation and Use of Utilisation Factors*, Technical Memorandum TM5
Civil Aviation Authority *Offshore Helicopter Landing Areas: Guidance on Standards*, CAP 437
Department of Energy *Guidance on the Design and Construction of Offshore Installations*, 1990

Chapter 11

American National Association of Corrosion Engineers (NACE) *Control of Pipeline Corrosion A. W. Peabody First Ed.* 1967
American National Association of Corrosion Engineers (NACE) RP-01-60 RP-05-72
BSI CP 1021: 1973 (1979) *Code of Practice for Cathodic Protection*
ERA Technology Ltd *A Code of Practice for the Safe Use of Electrical Equipment Underwater*
Morgan, J. H. *Cathodic Protection – Its Theory and Practice, First Ed.*, Leonard Hill Books Ltd, 1959

Chapter 12

Chatfield, Christopher *Statistics for Technology*, Penguin, 1970
Fisher, R. A. and Yates, F. *Statistical Tables for Biological, Agricultural and Medical Research* (6th edn), Longman, 1963
Green, A. E. and Bourne A. J. *Reliability Technology*, Wiley, 1972
Nieuwhof, G. W. E. *An Introduction to Fault Tree analysis with Emphasis on Failure Rate Evaluation – Microelectronics and Reliability*, Vol. 14, pp. 105–119, 1975

Chapter 13

Thorn EMI Instruments *A Simple Guide to Insulation and Continuity Testing*, 1984
BS 142: (*see list of standards below*)
BS 4999: Part 143: 1987 (*see list of standards below*)
BS 5227: 1984 (*see list of standards below*)

BS 5405 (*see list of standards below*)
BS 5730: 1979 (*see list of standards below*)

Codes of practice, standards and regulations

Codes of practice

American Petroleum Institute (API) *Design and Installation of Electrical Systems for Offshore Production Platforms* (1st edn), 1978

American Petroleum Institute (API) *Recommended Practice for the Protection against Ignition Arising out of Static, Lightning and Stray Currents*, 19

BS CP 1013: 1965 'Code of practice: earthing'

BS CP 1017: 1969 'Code of practice: distribution of electricity on a construction and building site'

Chamber of Shipping of the UK *Tanker Safety Code*

Department of Energy *Guidance on the Design and Construction of Offshore Installations*, 1990

Department of Energy *Offshore Installations: Guidance on Firefighting Equipment* (2nd edn in draft), 1991

ERA Technology Ltd *A Code of Practice for the Safe Use of Electrical Equipment Underwater*

Health and Safety Executive (UK) *Memorandum of Guidance on the Electricity at Work Regulations*, 1989

Institute of Petroleum (UK) Model Code of Safe Practice Part 1 *Electrical Safety Code*, 1965

Institution of Electrical Engineers *IEE Recommendations for the Electrical and Electronic Equipment of Mobile and Fixed Offshore Installations* (1st ed), 1983

Institution of Electrical Engineers *IEE Wiring Regulations* (16th ed), 1991. Note that these regulations are for general guidance only, as offshore installations are specifically excluded from their scope.

British Standards

BS 88: 1988 'Cartridge fuses for voltages up to and including 1000 V AC and 1500 V DC'

BS 142: 'Electrical protective relays'

    Part 0: 1982 'General introduction and list of parts'
    Part 1: 1982–9 'Information and requirements for all protection relays'
    Part 2: 1982–4 'Requirements for the principal families of protection relays'
    Part 3: 1983 'Requirements for single input energizing quantity relays
    Part 4: 1984 'Requirements for multi-input energizing quantity relays'

BS 923: 1980 'Guide on high voltage testing techniques'

BS 2692: 1986 'Fuses for voltages exceeding 1000 V AC'

BS 3871: (withdrawn) (see BS 4752) 'Moulded case and miniature circuit breakers up to 415 volts';

BS 3938: 1973 (1982) 'Current transformers'

BS 3939: Parts 1–13 (1985–6) 'Graphical symbols for electrical equipment'

BS 3941: 1975 (1982) 'Voltage transformers'

BS 3950: (withdrawn) 'Electrical protective systems for AC plant'

BS 4211: 1987 'Steel ladders for permanent access'

BS 4533 'Electric luminaires (lighting fittings)'
    Part 101: 1987 'General requirements and tests'
    Part 102: 1981–8 'Detail requirements Section 2.1: luminaires with type of protection "N" '

BS 4683 'Specification for electrical apparatus for explosive atmospheres'
    Part 1: 1971 'Classification of maximum surface temperatures' (IEC 79–8)
    Part 2: 1971 'The construction and testing of flameproof enclosures of electrical apparatus' (IEC 79–1)
    Part 3: (withdrawn: replaced by BS 6941: 1988) 'Type of protection "N" ' (IEC 79–1)
    Part 4: (obsolete: 1990) 'Type of protection "e"' (IEC 79–7)

BS 4727: Part 1: 1971–86 'Relay and measurement terminology'
    Part 2: 1971–86 'Terms particular to power engineering'

BS 4752: 'Circuit breakers of rated voltage up to and including 1000 volts AC and 1200 volts DC'
    Part 1: 1977 'Circuit breakers' (replaces BS 3871)

BS 4999 'Specification for general requirements for rotating electrical machines'
    Part 105: 1988 'Classification of degrees of protection provided by enclosures for rotating machinery'
    Part 143: 1987 'Tests'

BS 5209: 1975 (1980) 'Code of practice for the testing of sulphur hexafluoride taken from electrical equipment'

BS 5227: 1984 'AC metal-enclosed switchgear and control gear of rated voltage above 1 kV and up to and including 72.5 kV'

BS 5242 'Specification for contactors, oil, airbreak and vacuum'
    Part 1: 'Switchgear up to 1000 V AC'

BS 5266: 'Emergency lighting'
    Part 1: 1988 'Code of practice for emergency lighting of premises other than cinemas and certain other specified premises used for entertainment'
    Part 3: 1981 'Specification for small power relays (electromagnetic) for emergency lighting applications up to and including 32 A'

BS 5311: 1988 'AC circuit breakers of voltage above 1 kV'

BS 5345 'Code of practice for the selection, installation and maintenance of electrical apparatus for use in potentially explosive atmospheres (other than mining applications or explosive processing or manufacture)'
    Part 1: 1989 'Basic requirements of all parts of the code'
    Part 2: 1983 'Classification of hazardous areas'
    Part 3: 1979 'Installation and maintenance requirements for electrical apparatus with type of protection "d": flameproof enclosure'
    Part 4: 1977 'Installation and maintenance requirements for electrical apparatus with type of protection "i": intrinsically safe electrical apparatus and systems'

Part 5: 1983 'Installation and maintenance requirements for electrical apparatus protected by pressurisation "p" and by continuous dilution, and for pressurised rooms'

Part 6: 1978 'Installation and maintenance requirements for electrical apparatus with type of protection "e": increased safety'

Part 7: 1979 'Installation and maintenance requirements for electrical apparatus with type of protection "N": non-sparking'

Part 8: 1980 'Installation and maintenance requirements for electrical apparatus with type of protection "S": special protection'

BS 5405: (withdrawn: replaced by BS 6423: 1983 and BS 6626: 1985) 'Maintenance of electrical switchgear for voltages up to and including 145 kV'

BS 5420: 1977 (1988) Specification for Degrees of Protection of Enclosures of Switchgear and Controlgear for Voltages up to and Including 1000 V AC and 1200 V DC.

BS 5424: 'Specification for Controlgear for Voltages up to and including 1000 V AC and 1200 V DC.

Part 1: 1977 'Contactors'

Part 2: 1987 'Specification for Semiconductor contactors'

Part 3: 1988 'Additional requirements for contactors subject to certification'

BS 5486: 'Specification for factory built switchgear assemblies up to 1000 V AC'

Part 1: 1977 'General Requirements'

Part 1: 1986 'Specification for type-tested and partially type-tested assemblies (general requirements)'

Part 2: 1988 'Particular requirements for busbar trunking systems (busways)'

Part 11: 1989 'Specification for the particular requirements of fuseboards'

Part 12: 1989 'Specification for particular requirements of type-tested miniature circuit-breaker boards'

BS 5490: 1977 (1985) 'Specification and classification of degrees of protection provided by enclosure'

BS 5501: 'Electrical apparatus for potentially explosive atmospheres: Parts 1 to 9 inclusive:

Part 1: 1977 'General Requirements'

Part 3: 1977 'Pressurized apparatus 'p' '

Part 5: 1977 'Flameproof enclosure 'd' '

Part 6: 1977 'Increased safety 'e' '

Part 7: 1977 'Intrinsic safety 'i' '

Part 8: 1988 'Encapsulation 'm' '

Part 9: 1982 'Specification for intrinsically safe electrical systems 'i' '

BS 5730: 1979 'Code of practice for maintenance of insulating oil'

BS 5750: 'Quality Systems'

Part 0: 1987 'Principal Concepts and Applications'

Part 1: 1987 'Quality Assurance: Specification for Design, Manufacture and Installation'

Part 2: 1987 'Guide to the Use of BS 5750: Part 1'

Part 3: 1987 'Specification for final inspection and test'

Part 4: 1981 'Guide to the use of BS 5750: Part 1'

Part 5: 1981 'Guide to the use of BS 5750: Part 2'
Part 6: 1981 'Guide to the use of BS 5750: Part 3'
BS 6133: 1985 'Code of practice for the safe operation of lead-acid stationary cells and batteries'
BS 6346: 1989 Specification for PVC insulated cables for electricity supply'
BS 6423: 1983 'Code of Practice for maintenance of electrical switchgear and controlgear for voltages up to and including 650 V'
BS 6626: 1985 'Code of Practice for maintenance of electrical switchgear and controlgear for voltages above 650 V and up to and including 36 kV'
BS 6656: 1986 Guide to the prevention of inadvertent ignition of flammable atmospheres by radio-frequency radiation'
BS 6883: 1991 'Elastomer insulated cables for fixed wiring in ships and on mobile and fixed offshore units'
BS 6941: 1988 Specification for electrical apparatus for explosive atmospheres with type of protection N.

(Note: Standards are continually being revised and updated. Please refer to the current edition of the BSI Catalogue for the latest revisions.)

European standards

IEC 50: 'International electrotechnical vocabulary'
IEC 255: Parts 1–19 'IEC standards for electrical relays'
IEC 363: 1972 'Short-circuit current evaluation with special regard to related short-circuit capacity of circuit-breakers in installation in ships'

American standards

ANSI/IEEE Standard 43: 'Recommended practice for testing insulation resistance of rotating machinery'
C37–90: 'National standard for relays and relay systems'
National Fire Protection Association *Diesel-Electric Fire Pumps*, specification NFPA 20
American Petroleum Institute *Manual on Installation of Refinery Instruments and Controls. Part 1: Process Instrumentation and Control*, Recommended Practice RP 550
American National Association of Corrosion Engineers (NACE) RP-01-69 RP-05-72

UK Regulations

Mineral Workings (Offshore Installations) Act 1971, HMSO
The Health and Safety at Work Act 1974, HMSO
Offshore Installations (Construction and Survey) Regulations 1974, HMSO
Offshore Installations (Fire-Fighting Equipment) Regulations 1978, HMSO
The Electricity at Work Regulations 1989, HMSO

Index